A Colour Atlas of

Haematological Cytology

Second Edition

F.G.J. Hayhoe

MA, MD(Cantab), FRCP, FRCPath
Leukaemia Research Fund Professor of
Haematological Medicine, Cambridge University

R.J. Flemans

Consultant Technical Adviser
Department of Haematological Medicine
Cambridge University

Wolfe Medical Publications Ltd

Copyright © F.G.J. Hayhoe & R.J. Flemans, 1982
Second edition published by Wolfe Medical Publications Ltd, 1982
Printed by Royal Smeets Offset b.v., Weert, Netherlands
ISBN 0 7234 0778 9 cased edition
ISBN 0 7234 0985 4 limp edition
Reprinted 1987, 1990, 1991

A CIP catalogue record for this book is available from the
British Library.

This book is one of the titles in the series of Wolfe Medical
Atlases, a series that brings together the world's largest
systematic published collection of diagnostic colour
photographs.

For a full list of Atlases in the series, plus forthcoming titles
and details of our surgical, dental and veterinary Atlases,
please write to Wolfe Medical Publications Ltd, 2–16
Torrington Place, London WC1E 7LT, England.

Contents

Acknowledgements

We are grateful to our many colleagues, past and present in Cambridge, who have helped us to choose material for illustrations, and to numerous haematologists who have sent us unusual blood or marrow smears for opinions and thus added to our library of slides. We are especially indebted to Mrs J. M. Chandler for typing the manuscript and disentangling the index.

Introduction

Diagnostic advance since the first edition of this book was published in 1969 has not outdated the original illustrations, which have almost all been retained, although the colour rendering has been improved in many instances. It has, however, called for the inclusion of much new material, covering entities such as acute promyelocytic leukaemia, prolymphocytic leukaemia and hairy cell leukaemia, providing more extensive cytochemical illustration, especially of acid phosphatase and dual esterase reactions, and adding a whole section on imprint cytology of lymph nodes and spleen. The book now contains 750 colour photomicrographs in comparison with the 349 of the first edition.

We have continued to exclude histological material, whether of bone marrow or lymph node sections, because the interpretation of such material is generally the responsibility of the histopathologist rather than the haematologist.

Despite these additions the intent of the book remains the same. The second edition like the first is designed as an atlas, not a textbook. We assume that the user will have access to, and be potentially familiar with, the contents of standard works on theoretical and practical haematology and we do not feel a need to duplicate these expositions. We believe firmly that there is no substitute for the personal study of preparations under the microscope in gaining an understanding of haematological cytology, but we are equally convinced that a reference collection of photomicrographs in colour can provide a useful supplement to such microscopic study. Good atlases of haematology exist, but there remains a need for a relatively inexpensive volume of high quality photomicrographs which would be suitable for day-to-day use at the bench. This is what we have aimed to provide. The photographs show examples of most cell types likely to be encountered at all commonly and of some which are less common, but we have avoided disproportionate illustration of rarities. There has been no selection of superbly photogenic fields, as our objective has been verisimilitude rather than perfection. We judge the variability in staining and in background among the collection of photographs not to be inimical to practical use but helpful in aiding appreciation of the tinctorial variation implicit in most staining methods in common laboratory use. Most of the photographs are from smears stained with the Leishman or May-Grunwald-Giemsa methods, but examples of cytochemical staining with methods that have acquired a routine diagnostic value are included. The methods in question are the Prussian blue reaction for free iron, Sudan black B for lipids, the periodic acid-Schiff (PAS) reaction for glycogen (chiefly), and techniques for peroxidase, acid and alkaline phosphatases and for chloroacetate and butyrate (or acetate) esterases. Technical details of these methods are given in an Appendix.

The work is divided into five sections: the red cells and their precursors; non-erythroid cells of myeloid origin (granulocytes, monocytes, megakaryocytes and platelets); lymphocytes, plasma cells and their derivatives and precursors; miscellaneous cells of blood and marrow including foreign cells and parasites, and a fifth section illustrating the imprint cytology of normal and pathological lymph nodes and spleen, together with some examples of other biopsy material which might be confused with lymph nodes; this section also includes examples of cells found in pleural, ascitic and meningeal fluids that are of particular interest and importance to the haematologist, as requiring discrimination from leukaemic or lymphoma cells. Each section is preceded by a very brief introduction, somewhat longer in the case of the first section because the variants among red cells and their precursors are more copiously and confusingly categorised than is the case in the other cell series.

The descriptive legends accompanying the illustrations serve chiefly to identify the cells; the cytological data required for recognition of cell types is contained in the photographs, and for the intelligent and perceptive student there is no need to repeat in the text what the illustrations adequately convey. Some additional explanation of certain unusual appearances and of the cytochemical reactions is added where it seems necessary.

Photographs were taken with either Zeiss or Leitz photomicroscopes equipped with automatic exposure devices, using various 35 mm Kodak films with appropriate filtration to the correct colour temperature.

The specific magnification for each photograph is not stated; red cells are present in most fields depicted and provide a rough scale, while variations in the degree of spreading of films make comparisons between neighbouring cells a more useful index of cell size for the practical haematologist than an absolute scale.

Part 1

The red cells and their precursors
Normal forms and abnormal variants

The nomenclature of red cell precursors is confusing. The earliest recognisable member of the red cell series, the proerythroblast, has the cytoplasmic basophilia, the nucleolated and moderately leptochromatic nucleus and the large cell size generally characteristic of primitive cells. It gives rise to a sequence of nucleated cells, the erythroblasts, which progressively develop increasingly pachychromatic nuclei, lose their nucleoli and their cytoplasmic basophilia and acquire a rising haemoglobin content. This sequence is subject to an arbitrary division into stages, the commonest division being into three:

(1) The basophilic or early erythroblast, or normoblast A
(2) The polychromatic or intermediate erythroblast, or normoblast B
(3) The orthochromatic or late erythroblast, or normoblast, or normoblast C.

There are objections to the use of many of these terms, but they are all so firmly entrenched in common usage that they must be accepted. When authors use different or more elaborate staging and nomenclature they usually define their terminology, but those who use any of the synonyms above expect them to be understood without further explanation.

The proerythroblast is not itself the functional stem cell serving as a self-maintaining progenitor of the normoblast series, but is derived from an earlier functional myeloid stem cell of unidentified morphology having pluripotential capacity for giving rise to cells of erythroid, granulocytic, monocytic and megakaryocyte-platelet lines.

Kinetic studies with radio-isotopically labelled cells suggest that four cell cycles culminating in mitoses occur during development from proerythroblast to late normoblast, three at the proerythroblast and early basophilic normoblast stages, and the last at the polychromatic intermediate normoblast stage. Nests of erythroblasts of different stages of maturity commonly occur in apposition around a centrally situated reticulo-endothelial cell. Transfer of iron may be effected in one or other direction, and the central macrophage is often rich in stainable free iron. Late normoblasts do not undergo a further cell cycle but lose their nuclei by extrusion and give rise to marrow reticulocytes.

Reticulocytes spend up to two days in the bone-marrow before being released into the peripheral blood. There they make up normally less than 1 per cent of the red cell population and within another one to two days lose the remnants of cytoplasmic basophilia which give them their characteristic staining properties, and become orthochromatic mature red cells.

Mature red cells survive some 120 days before destruction. They are normally circular and fairly uniform in diameter, but are readily distorted by external pressure, as from neighbouring cells in a smear. Their structure as biconcave discs leads to weaker eosinophil staining at the centre than at the periphery, a feature which is least prominent at the tail of a blood smear where the cells are most spread out and flattened. In the body of the smear it becomes more conspicuous and this normal phenomenon must be appreciated and distinguished from hypochromia.

Abnormal variants
Nucleated precursors

The chief cytological variants involving alteration in morphology rather than numbers or relative proportions of erythroblasts are as follows:

Macronormoblasts: Cells having the general nuclear and cytoplasmic morphology of normoblasts but an increased average cell diameter. They occur especially in states of active erythroblastic hyperplasia without haematinic defects, such as haemolytic anaemias, but may be seen also in the early stages of disorders that subsequently become megaloblastic.

Megaloblasts: The essential morphological change in megaloblasts as compared with normoblasts lies in the more open chromatin pattern of the nucleus at all stages of development. There is often a degree of haemoglobinisation of the cytoplasm which appears excessive for the state of nuclear maturation, and later megaloblasts may appear fully orthochromic. Megaloblasts are larger than normoblasts of comparable maturity. They may show mitotic irregularities, with multipolar mitoses, sometimes asymmetrical, producing unequal daughter cells. Cytokinesis may not occur and giant cells with two or more nuclei, sometimes unequal in size, result. Chromosomes or fragments of chromatin may become separated from the spindle and constitute accessory small nuclear masses. These 'Howell-Jolly bodies' may remain in the red cells after the main bulk of nuclear material has been extruded. Rosette formation, with several chromatin masses linked by bridges, may follow a halt of the mitotic process in metaphase.

Megaloblasts are present in the marrow in states of B_{12} and folic acid deficiency, the commonest of which are Addisonian pernicious anaemia and conditions of gastro-intestinal malabsorption due either to disease or to gastric or intestinal resections. The folic acid deficiency of pregnancy may lead to megaloblastic anaemia. Megaloblasts are seen also in some cases of refractory sideroblastic anaemia and in erythraemic myelosis. They may be seen in acute leukaemias, especially following treatment with antimetabolites.

Micronormoblasts: Small normoblasts with a tendency to ragged cytoplasmic outline and irregular staining are found in states of iron deficiency. The changes are most evident in intermediate and late normoblasts, which also show defective haemoglobinisation.

Sideroblasts: Normoblasts containing free, non-haemoglobin, iron are best detected by the use of Prussian-blue staining, but occasionally coarse accumulations of free iron may be visible in Romanowsky preparations.

Bizarre cytological variants: Gross mitotic and nuclear abnormalities, of the same kind as described under megaloblasts but much more extreme in degree, occur in erythraemic myelosis.

Reticulocytes

Blood normally contains less than 1% of reticulocytes and an increase in numbers indicates a heightened output of young red cells from the bone marrow. This may occur during the correction by increased marrow activity of any anaemic state, whether spontaneously (as after an acute haemorrhage) or following treatment (as in pernicious anaemia after B_{12} or iron deficiency anaemia after iron). Reticulocytosis is especially prominent in haemolytic anaemias, where great erythropoietic activity attempts to compensate for the shortened life span of peripheral erythrocytes.

In Romanowsky preparations reticulocytes may sometimes be detected as polychromatic cells, with a blue or purple tinge, or much less often may show a scattering of fine blue 'basophilic stipples'. They are best recognised, however, by the use of supravital staining with brilliant cresyl blue, which reveals a network of fine filaments and dots. Reticulocytes are often a little larger in diameter than mature red cells.

Mature red cells

Red cells may differ from normal in their appearance in stained films by variations in size (anisocytosis, microcytosis, macrocytosis), in shape (poikilocytosis, elliptocytosis, sickle cell formation, crescent formation, crenation) and in depth of staining (hypochromia, anisochromia, spherocytosis, target cell formation). They may also undergo fragmentation (schistocytosis) or contain abnormal inclusions, such as residual chromatin material – Howell-Jolly bodies.

All these abnormalities may be seen in varying degree and frequency and several of them may occur together in the same smear.

1. Sequence from proerythroblast through early, intermediate and late normoblasts to non-nucleated red cells.

The gradual progression in nuclear and cytoplasmic maturation shown here indicates the artificial nature of the arbitrary division into stages.

2. A proerythroblast and six intermediate to late normoblasts.

3. A proerythroblast and two early normoblasts.

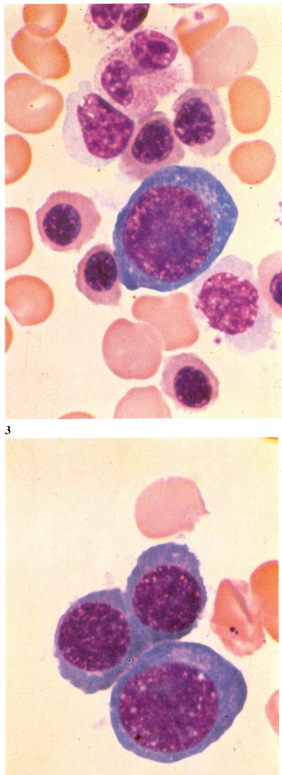

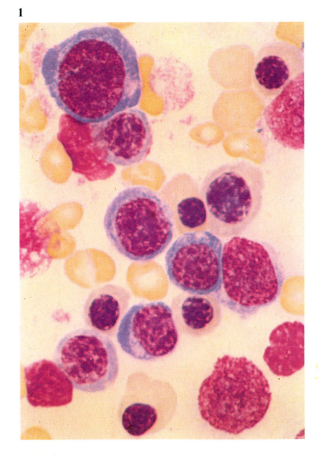

4. A proerythroblast in early prophase of mitosis.

5. A proerythroblast, late normoblast and an early normoblast or proerythroblast in mitosis (metaphase).

6. A group of normoblasts with accompanying granulocytes for comparison.

5

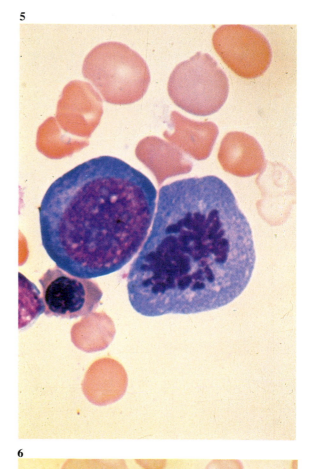

4

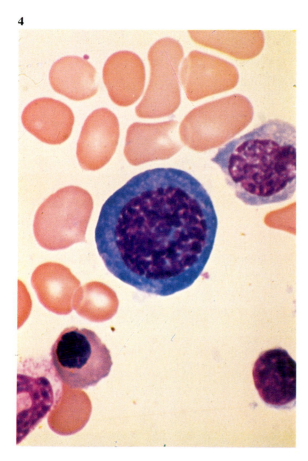

6

7. A group of normoblasts with two granulocyte precursors and a plasma cell, for comparison.

8 and 9. Nests of normoblasts around a central macrophage.

8

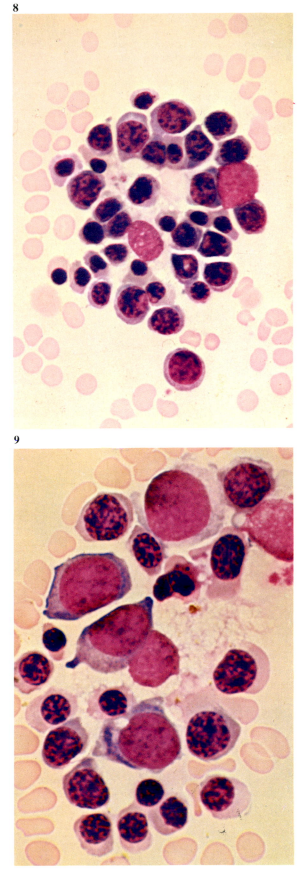

7

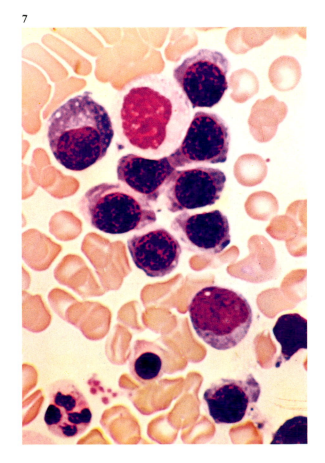

9

10. Normoblasts around a macrophage – alkaline phosphatase reaction. The strong positivity of the macrophage spills over to give threads of positive reaction over the surface of contiguous normoblasts.

11 and 12. Nests of normoblasts around a central macrophage – Prussian blue stain for free iron. The iron-containing cytoplasm of the macrophage extends as a long broad tail across the field in **12**.

11

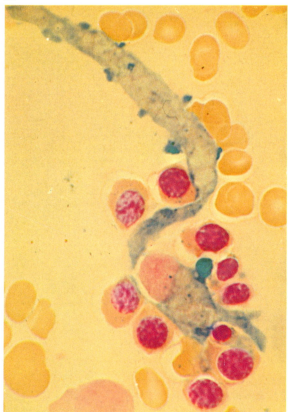

10

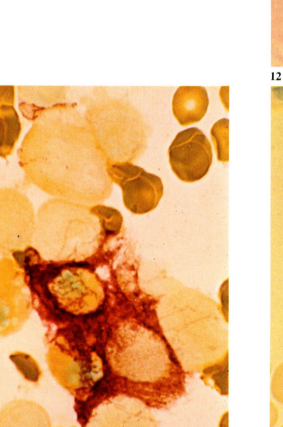

12

13. Normal mature red cells, from the tail of a blood smear. Minor variations in size and depth of staining are seen, together with occasional pressure distortion of shape.

14. Normal mature red cells from the body of a blood smear. There is a greater tendency here for cells to overlap one another, and the weaker staining of the central area is more evident.

15. Mixture of normal adult and cord blood red cells – Kleihauer reaction. The HbF-containing cord cells remain unlysed, while the HbA-containing adult cells show haemolysis. Acid and alkali resistance less marked than that of cord cells but greater than that of normal red cells may be shown by the red cells in hereditary persistence of HbF and in some thalassaemias.

14
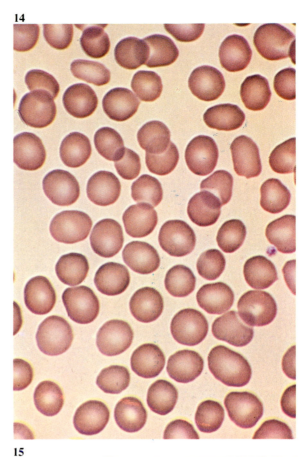

13
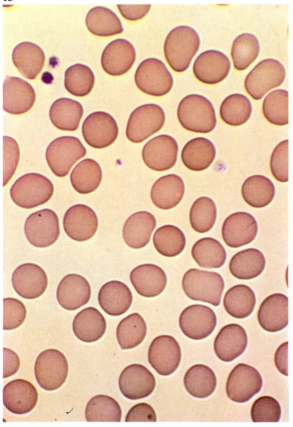

15
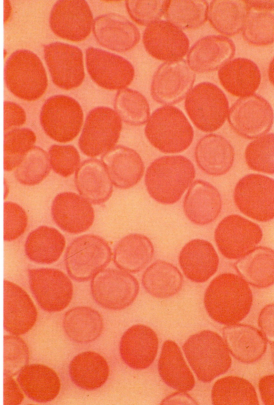

16. A group of early basophilic macronormoblasts.

17. Early and intermediate macronormoblasts, showing cytoplasmic maturation with accumulation of haemoglobin in some cells to a degree more advanced than would normally accompany their state of nuclear maturation.

18. Macronormoblasts, from early to late stages. There is a lymphocyte at the top.

17

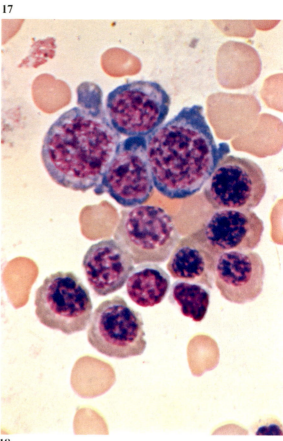

16

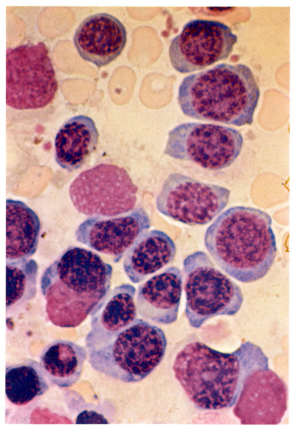

18

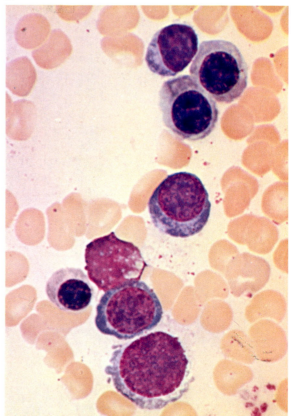

19. Late telophase of mitotic division in a macronormoblast, with daughter cells each containing two unequal nuclear masses. A mature neutrophil and a myelocyte are also present.

20. Macronormoblastic hyperplasia in foetal haemolytic disease, stained by the PAS reaction. Normal erythroblasts are invariably PAS negative, but positive material (probably glycogen, and removable by treatment with salivary amylase) is found in certain pathological states, including especially foetal haemolytic disease, thalassaemia, iron deficiency anaemia and erythraemic myelosis or erythroleukaemia.

Here two late erythroblasts show moderately heavy positive stippling.

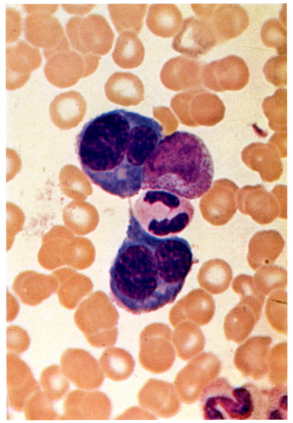

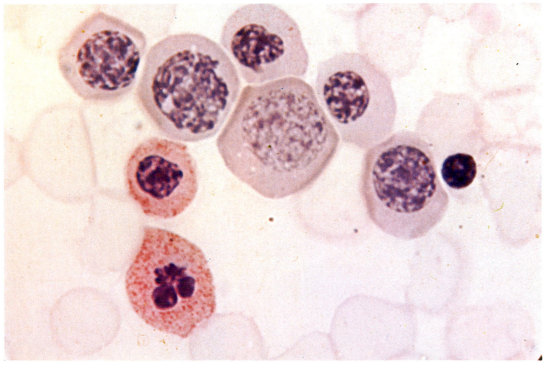

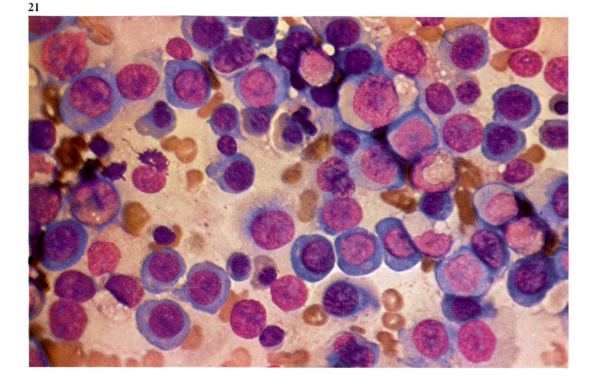

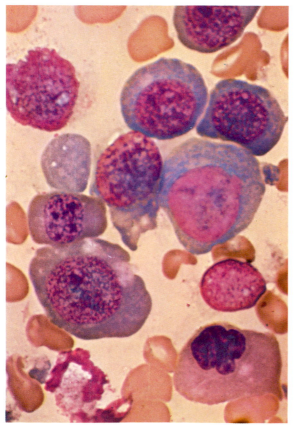

21. A general view of a marrow smear from a patient with pernicious anaemia. Erythroblasts greatly predominate, and erythropoiesis is megaloblastic. Early stages in the sequence from proerythroblast onwards are especially common.

22. Proerythroblasts (nucleolated) and early and intermediate megaloblasts. While the nuclear pattern of proerythroblasts in pernicious anaemia and other megaloblastic anaemias is not distinctively different from that of normal proerythroblasts, there is a tendency for cytoplasm to be more abundant and nucleoli larger and more conspicuous. The field contains a late megaloblast of large size and having an irregular pycnotic nucleus. There is also a separated fragment of polychromatic megaloblast cytoplasm.

23 and 24. Gross megaloblastic changes resulting from the action of antimetabolites (cytosine arabinoside and 6-thioguanine) in a patient with acute myeloid leukaemia now moving into remission.

25. Two proerythroblasts and three early megaloblasts, showing variability in size and cytoplasmic staining. Basophilia is often, as here, deeper in the early megaloblast cytoplasm than in the proerythroblasts.

24

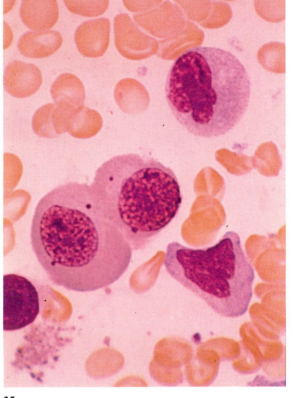

23

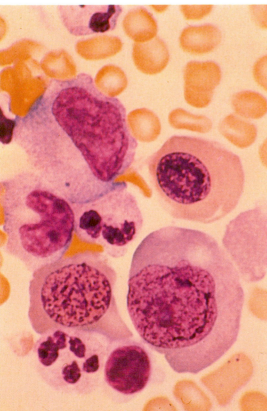

25

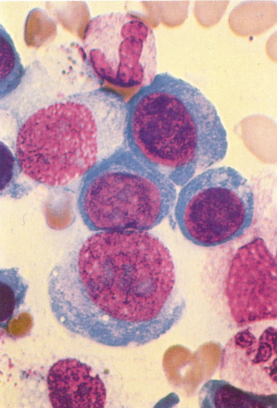

26. An early megaloblast with nucleolar traces and deep cytoplasmic basophilia, and an intermediate megaloblast of gigantic size.

27. A mature neutrophil with a proerythroblast, an early to intermediate megaloblast and a late megaloblast. This last cell has not quite reached the stage of complete nuclear pycnosis, but contains a separated chromatin mass, lost from the spindle at a preceding mitosis. This is a Howell-Jolly body.

28. Early, intermediate and late megaloblasts, the last containing two Howell-Jolly bodies.

27
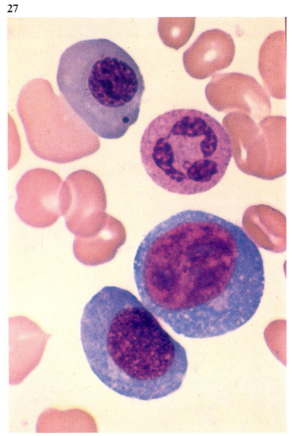

26
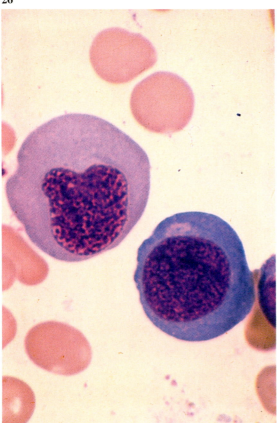

28
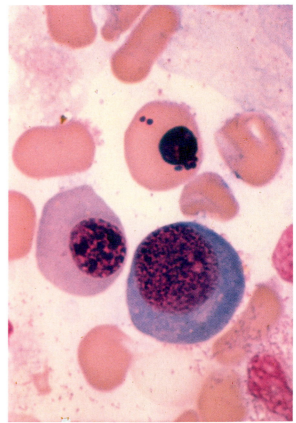

29. A gigantic early megaloblast with four nuclei, probably resulting from two consecutive incomplete mitoses, with nuclear division unaccompanied by cytoplasmic division.

30. An intermediate and a late megaloblast. The former shows nuclear distortion of minor degree.

31. An intermediate megaloblast with two nuclei.

32. Late megaloblasts. One is gigantic, with two connected irregular nuclear masses and a free Howell-Jolly body.

33. An irregular nuclear mass approaching extrusion in a late megaloblast.

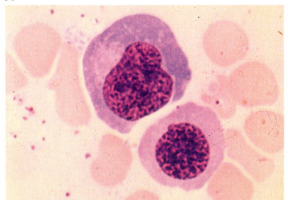

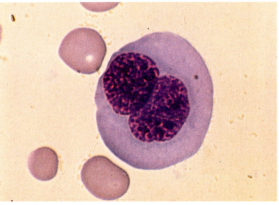

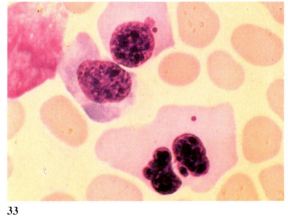

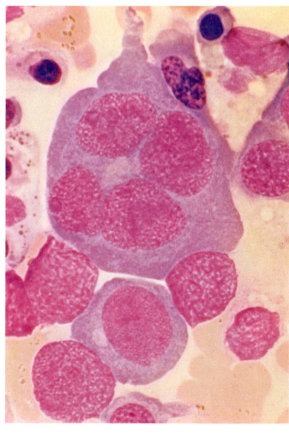

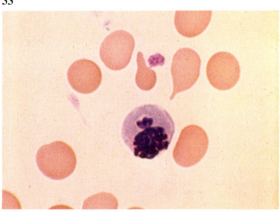

34 and 35. Late megaloblasts showing typical nuclear rosette formation, probably arising from an incomplete mitosis with hold-up at metaphase.

36. Metaphase of mitosis in an early megaloblast. The chromosomes are notably long and thin and unusually widely scattered.

37. Telophase in late megaloblasts. The mitotic process has been defective and several chromatin fragments (perhaps whole chromatids) have been lost from the spindle and appear separate from the main bodies of the reconstituted nuclei. This illustrates the genesis of Howell-Jolly bodies.

38. An abnormal tripolar mitosis, approaching telophase.

36

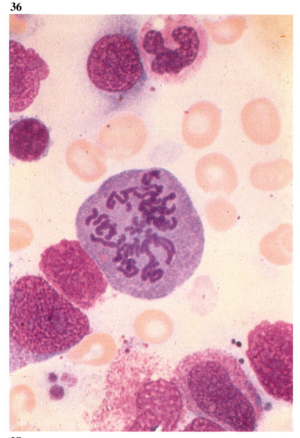

34

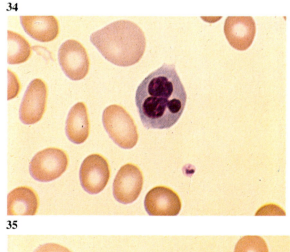

35

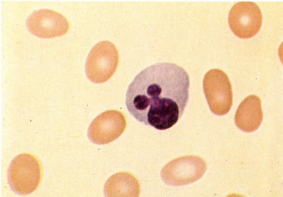

37

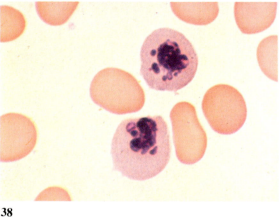

38

39. An intermediate megaloblast and a lymphocyte (for comparison) in the blood.

40. An intermediate megaloblast with nuclear distortion. The eccentric position of the nucleus and the persistent basophilia of the cytoplasm gives a resemblance to a plasma cell.

41 and 42. Late megaloblasts in the peripheral blood. Each shows almost complete extrusion of the pycnotic nucleus. With loss of the nucleus a polychromatic reticulocyte or young erythrocyte will remain.

43. A proerythroblast and a late megaloblast, surrounded by erythrocytes showing macrocytosis and minor aniso- and poikilocytosis. A few small Howell-Jolly bodies are visible.

41

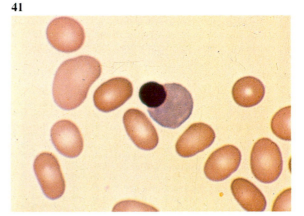

42

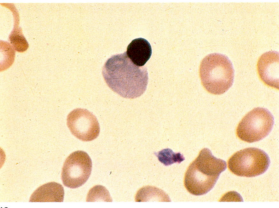

39

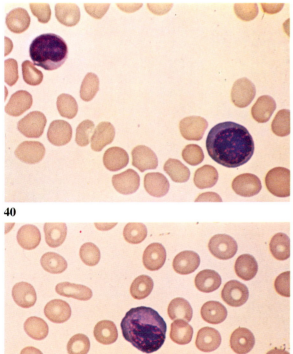

40

43

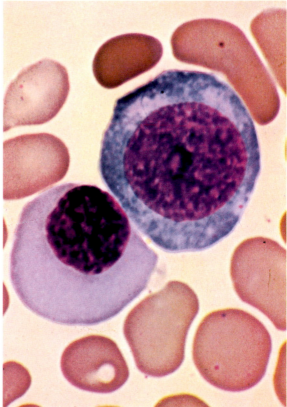

44. A range of erythroblasts with minimal megaloblastic change. The confusing terms 'intermediate' or 'transitional' megaloblasts are sometimes applied to such erythroblasts, whether early or late in the maturation sequence.

45 and 46. Further groups of erythroblasts with minor megaloblastic changes. The nuclear pattern is more open than that of comparable normoblasts, but less so than in florid megaloblasts. The cell at the bottom of **45** is a lymphocyte, and the loss of nucleus from a late normoblast at the top leaves a stippled red cell.

45

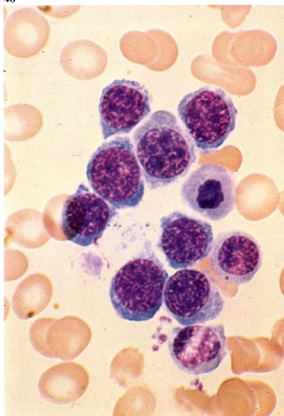

44

46

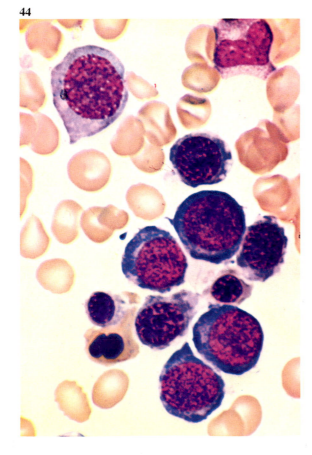

47–49. A series of fields illustrating recognizable but minor megaloblastic changes in erythroblasts from the bone marrow of a patient in an early stage of pernicious anaemia.

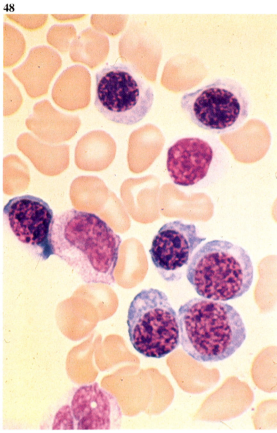

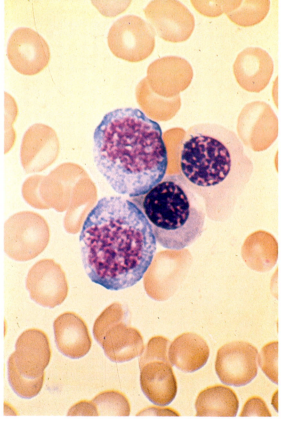

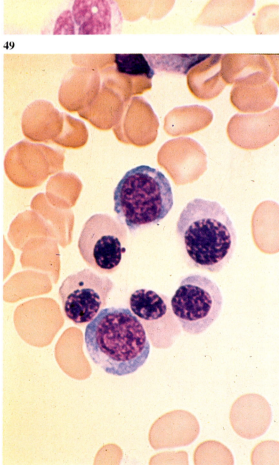

50. Further minor megaloblastic changes in erythroblasts. The large cell in the centre is a 'giant metamyelocyte' characteristically found in pernicious anaemia.

51. Polychromasia in a Leishman-stained preparation from the peripheral blood of a patient with a megaloblastic anaemia responding to specific treatment. Such polychromatic erythrocytes tend, as here, to be larger than more mature orthochromatic cells, and with cresyl blue supravital staining are shown to be reticulocytes.

52. A similar specimen stained with cresyl blue supravitally. Note that the haemoglobin content of the reticulocytes appears high – an element of hypochromia accompanied the megaloblastic changes in this case. 'Dimorphic' patterns with macrocytosis and hypochromia and deficiencies of both vitamin B_{12} (or folic acid) and iron occur especially in association with intestinal malabsorption.

51

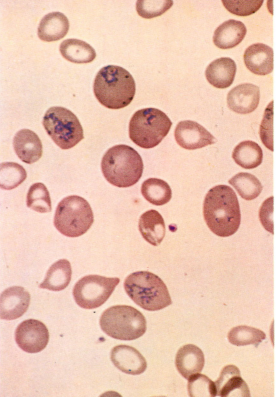

50

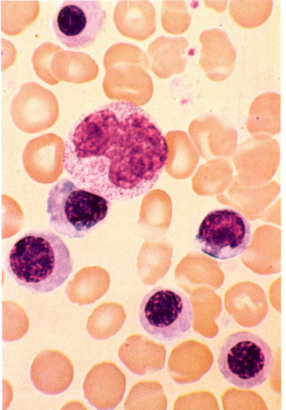

52

53. Peripheral blood in pernicious anaemia. There are two late megaloblasts showing nuclear rosette formation and basophil stippling (another manifestation of reticulocyte material). The red cells show macrocytosis, anisocytosis and poikilocytosis.

54. Similar changes in red cells from another area in the same specimen. This field includes a late megaloblast and a multi-lobed neutrophil polymorph.

55. Another field showing macrocytosis, and minor aniso- and poikilocytosis, with a striking multi-lobed neutrophil polymorph.

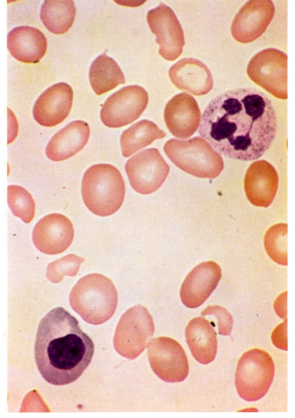

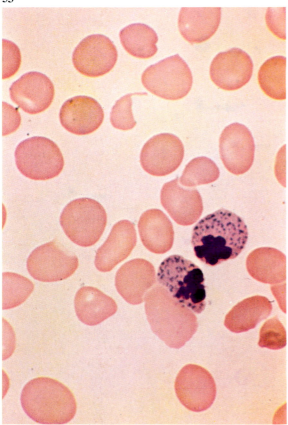

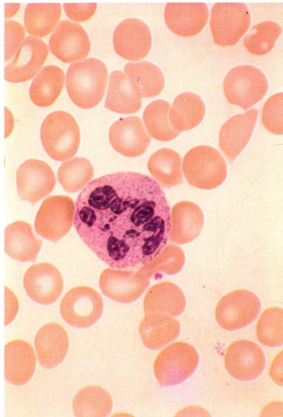

56. Vacuolation in proerythroblasts following toxic reaction to chloramphenicol. The proerythroblasts are Sudan black negative, while the sudanophilic granulocytes show no vacuolation.

57 and 58. Examples of giant proerythroblasts in the bone marrow of a patient with hereditary spherocytosis during the early stages of recovery from an aplastic crisis.

57

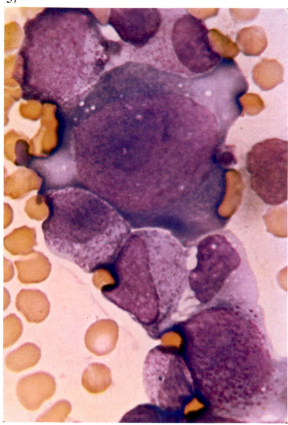

56

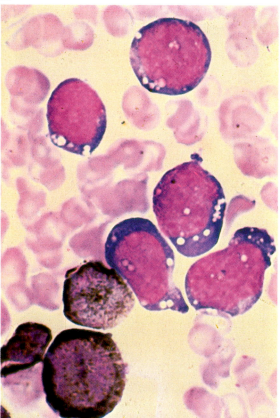

58

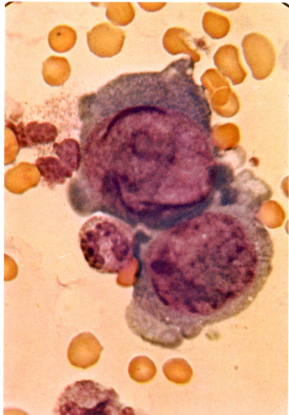

59 and 60. Congenital dyserythropoietic anaemia (CDA) type I. Later basophilic and polychromatic erythroblasts chiefly affected; megaloblastoid changes; binucleated cells with internuclear chromatin bridges; spongy chromatin with irregular nuclear outline.

61. CDA type II. Late erythroblasts chiefly affected; many binucleated cells. A double nuclear membrane is visible only on electron microscopy. In type II CDA (hereditary erythroblastic multinuclearity with positive acidified serum test – HEMPAS) the red cells are susceptible to acid haemolysis.

60

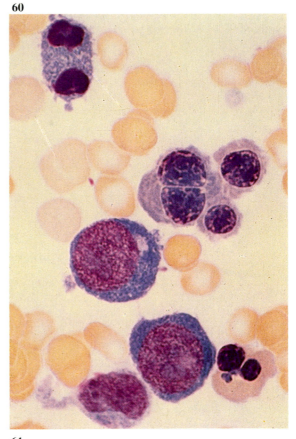

59

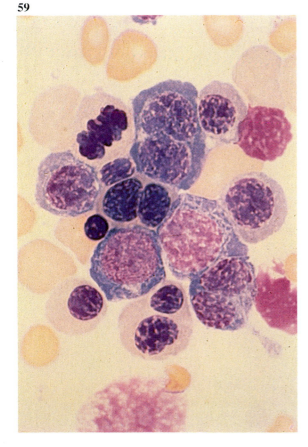

61

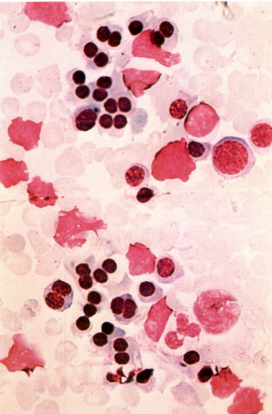

62. CDA type II. A further example showing two erythroblasts with double nuclei and one with four unequal nuclei.

63 and 64. CDA type III. Giant erythroblasts with multiple nuclei or a single large lobulated nucleus predominate.

63

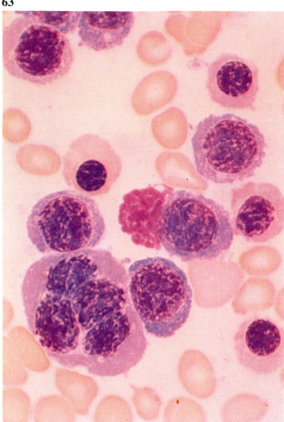

62

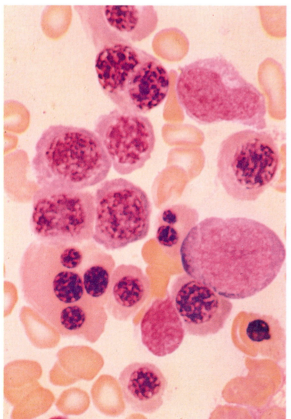

64

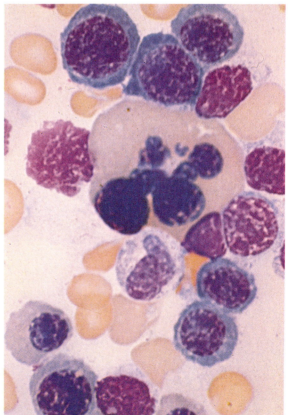

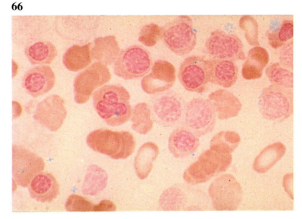

65–69. *Cytochemical reactions of erythroblasts in CDA. The patterns illustrated appear common to all types.*

65. PAS reaction shows weak diffuse and some finely granular positivity, much less than usual in erythraemic erythroblasts.

66. Prussian blue stain. Excess free iron in erythroblasts, but no ringed sideroblasts seen.

67. Kleihauer reaction. CDA erythroblasts and some erythrocytes show acid-resistant HbF.

68. Acid phosphatase: normal positivity in a macrophage and an eosinophil myelocyte. CDA erythroblasts are essentially negative, in contrast to erythraemic erythroblasts.

69. Double esterase: normal chloroacetate esterase (CE) positivity in a neutrophil stab cell: negative reaction in CDA erythroblasts, unlike the positive butyrate esterase (BE) in erythraemic erythroblasts.

67

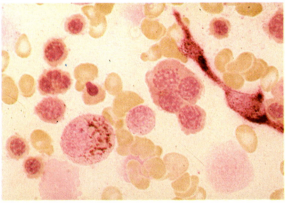

65

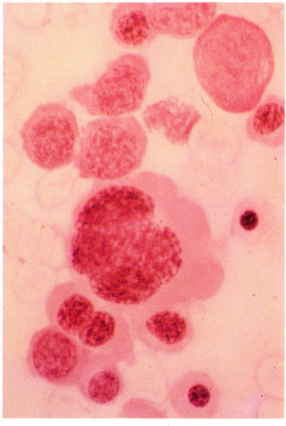

68

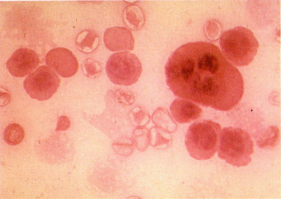

69

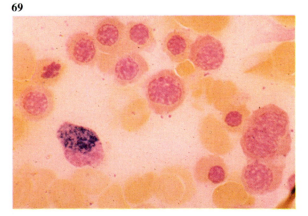

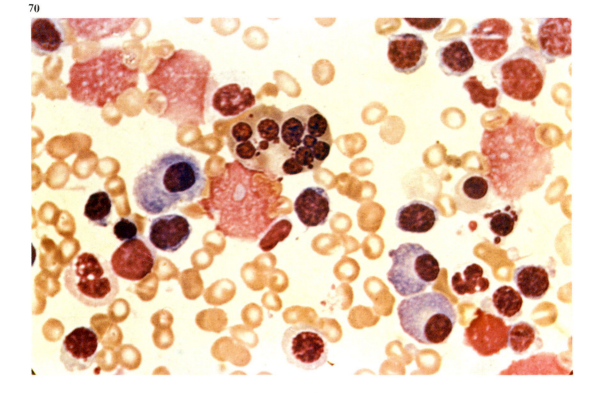

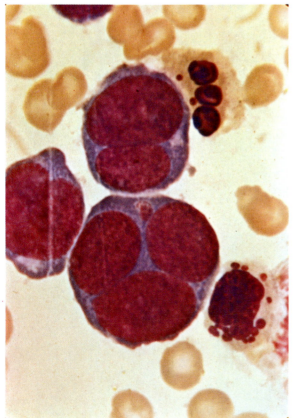

70. Bizarre nuclear abnormalities in erythroblasts from the bone marrow of a patient with erythraemic myelosis. Most cells present are erythroblasts (apart from three plasma cells) and several show multiple nuclear masses of irregular size and shape, as well as smaller Howell-Jolly bodies.

71. Multinucleated giant erythroblasts or proerythroblasts in erythraemic myelosis.

72–75. Highly abnormal erythroblasts from erythraemic myelosis.

Cellular gigantism, multiple and irregular nuclear masses and abnormal mitotic figures occur in erythroblasts at different stages of nuclear maturation and cytoplasmic haemoglobinisation.

Abnormalities as striking and bizarre as these are virtually restricted to erythraemic myelosis and erythroleukaemias, but they are probably not essentially different in kind from the similar but much less frequent and prominent nuclear abnormalities illustrated earlier as occurring in the common megaloblastic anaemias, or the frequent but more regular multinuclearity of CDA.

73

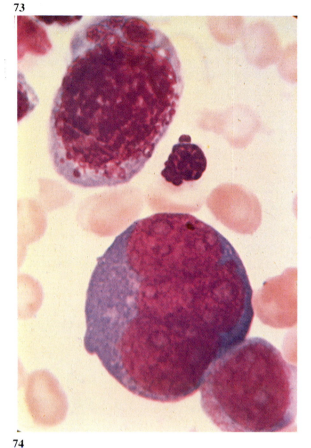

72

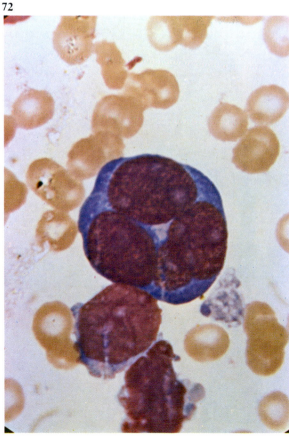

74

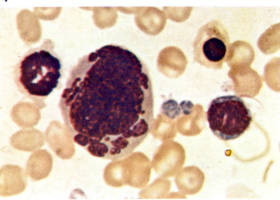

75

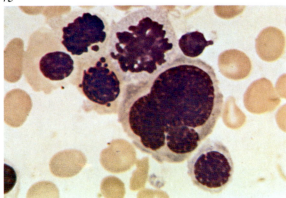

76. A group of proerythroblasts with some later erythroblasts of various stages of maturity surrounding them, from a patient with erythraemic myelosis. Mitotic abnormalities were not gross or frequent in this case, but erythroblasts vastly predominated.

77. Another example of bone marrow cytology in erythraemic myelosis. Although the cells are mostly proerythroblasts, they show a suggestion of megaloblastic change. Erythroblasts are always negative to Sudan black and for peroxidase. This smear shows a peroxidase stain, with strong positivity in two neutrophils but complete absence of reaction in the red cell precursors.

78. Sudan black stain on a further marrow aspirate from erythraemic myelosis. The erythroblasts are negative while three granulocyte precursors show characteristic positivity.

77

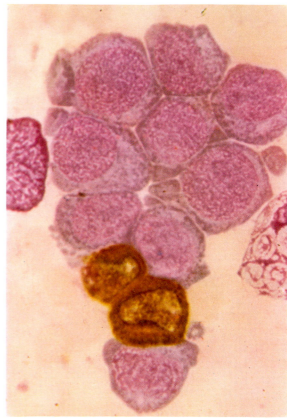

76

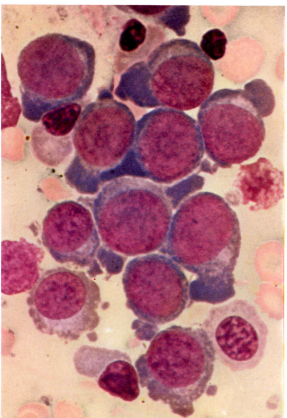

78

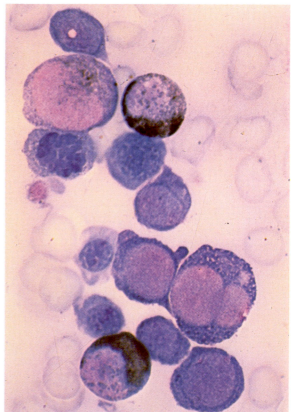

79 and 80. Two examples of PAS positivity in erythroblasts from erythraemic myelosis. Coarse granular positivity is conspicuous in early erythroblasts, with diffuse positivity, sometimes very intense, in later erythroblasts.

81. PAS positivity in two grossly abnormal giant erythroblasts from erythraemic myelosis.

80

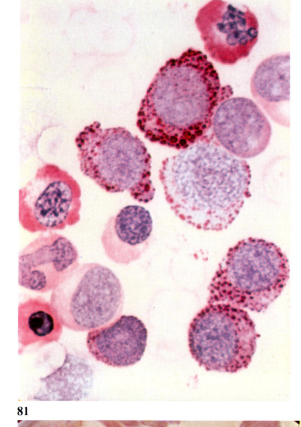

79

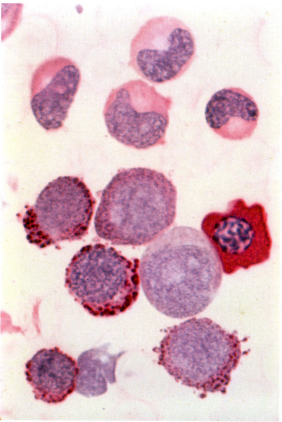

81

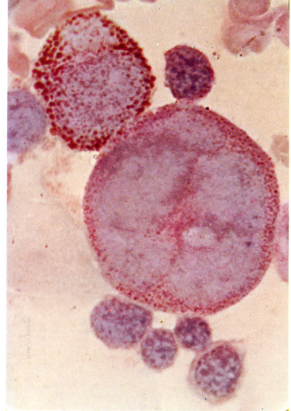

82 and 83. Further examples of PAS positivity in erythraemic erythroblasts, one in mitosis. Early granulocyte precursors in these two fields show negative reactions.

84. Acid phosphatase in erythraemic myelosis.

85. Another case of erythraemic myelosis showing even coarser acid phosphatase positivity in erythroblasts.

86. Double esterase in erythraemic myelosis. The early erythroblasts show a mixture of both BE and CE positivity, chiefly the former.

84

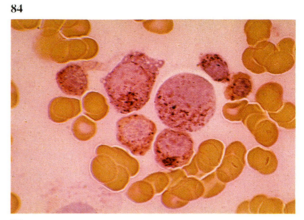

85

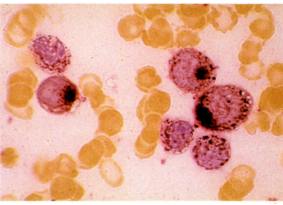

82

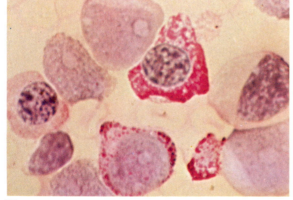

83

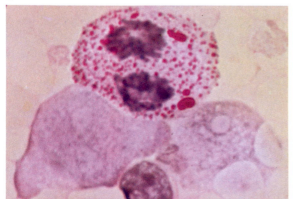

86

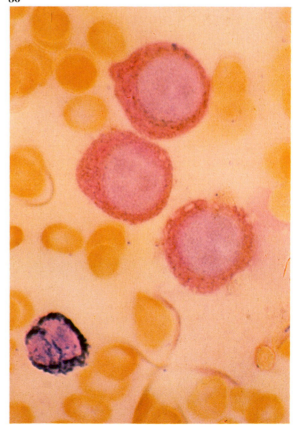

87 and 88. Erythroblasts from the bone marrow in iron deficiency anaemia.

Erythropoeisis is normoblastic, but the normoblasts tend to be small, with ragged outlines and defective haemoglobinisation.

89. PAS positivity in the small and ragged late erythroblasts of iron deficiency anaemia. This is a frequent but not invariable finding in iron deficiency.

88

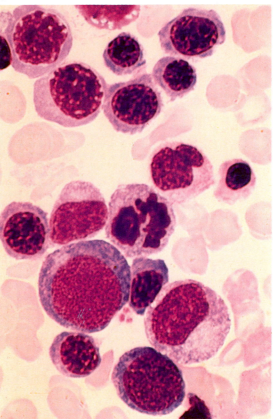

87

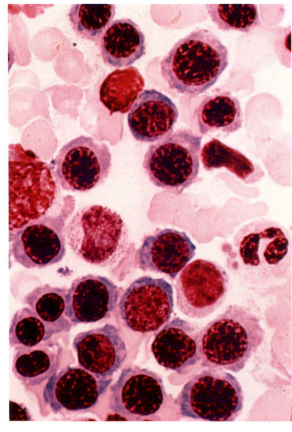

89

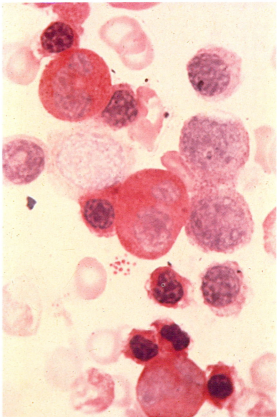

90. Leishman stain of a marrow fleck containing scattered brownish-black granules, probably mostly within RE cells.

91. Consecutive free iron stain on the same field showing that the granules react for iron.

92. A cellular 'fleck' from a bone marrow smear in iron deficiency anaemia, stained for free iron by the Prussian blue method. Free iron is absent.

93. A similar preparation from a patient with a sidero-blastic anaemia and increased iron stores. (Normal subjects have stainable free iron in amounts midway between these two extremes.)

92

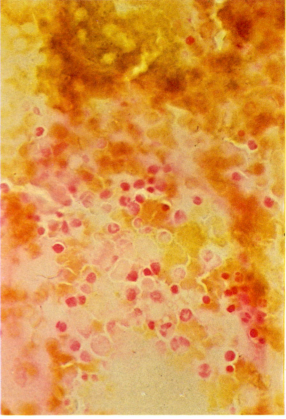

90

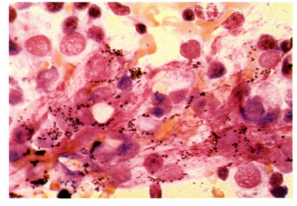

91

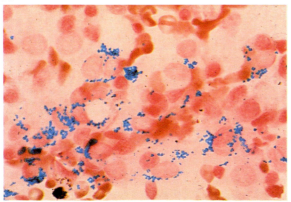

93

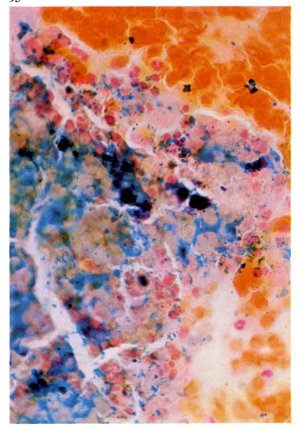

94. Leishman stain, bone marrow from a patient with sideroblastic anaemia. There is a suggestion of megaloblastic nuclear changes and irregular cytoplasmic staining in the erythroblasts.

95. A stain for free iron on the same marrow specimen, showing coarsely positive ringed sideroblasts – erythroblasts with free iron granules arranged as a continuous, or almost continuous, ring around the nucleus. This iron is chiefly concentrated in mitochondria.

96 and 97. Prussian blue stain for free iron showing further examples of ringed sideroblasts.

95

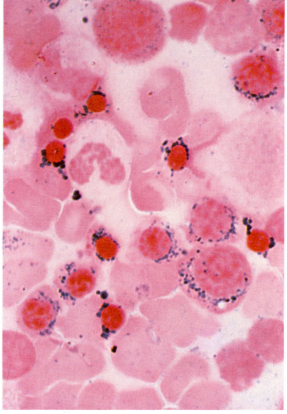

94

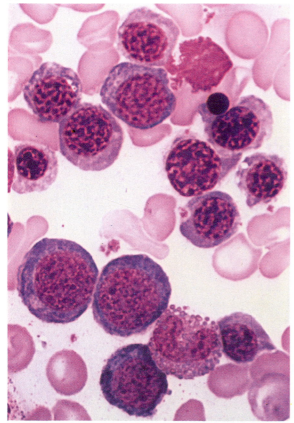

96

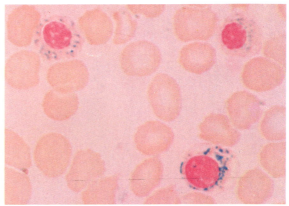

97

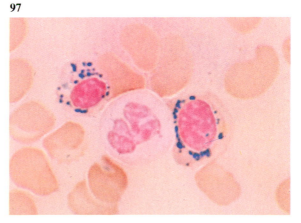

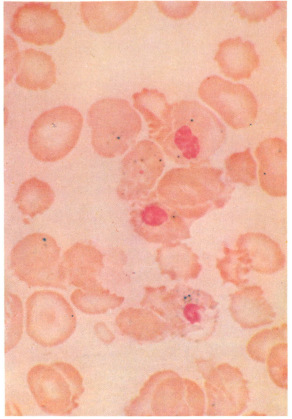

98. Free iron in erythroblasts (sideroblasts) with a non-ringed distribution. This example is of the 'normal' pattern of free iron distribution in erythroblasts, but the positive cells (sideroblasts) are more numerous than usual.

99. Consecutive reactions for PAS positivity and free iron in the marrow from a patient with refractory sideroblastic anaemia. There are ringed sideroblasts present and also granular PAS positivity in early erythroblasts.

Similar pictures may be seen in erythraemic myelosis, except that the iron tends to be less abundant and rarely in ringed form, and the PAS positivity is often stronger especially in later erythroblasts.

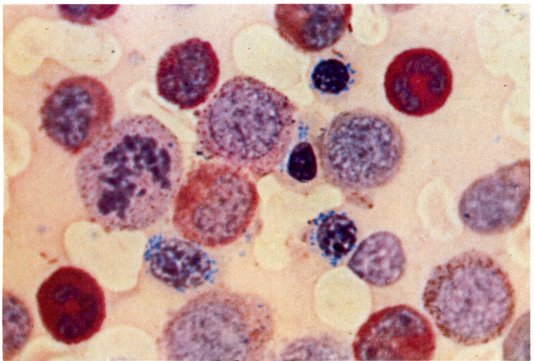

100–103. *Various appearances of reticulocytes.*

100. Polychromasia.

101. Basophil stippling.

102. Reticulin material after supravital brilliant cresyl blue staining with Leishman counterstain.

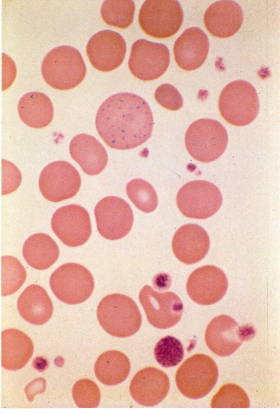

101

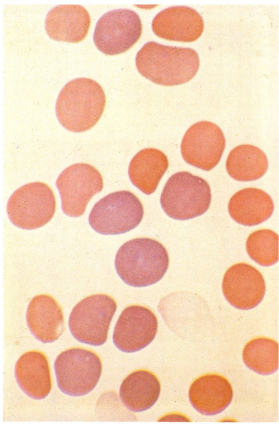

100

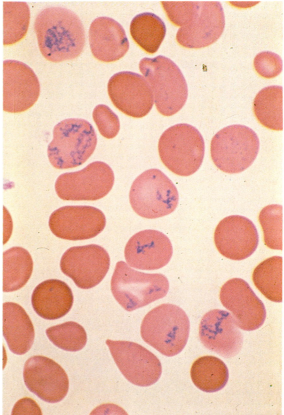

102

103. Reticulocytes stained supravitally, but uncounterstained.

104. Gross reticulocytosis – approaching 80 per cent – in a patient with haemolytic anaemia due to congenital pyruvate-kinase deficiency.

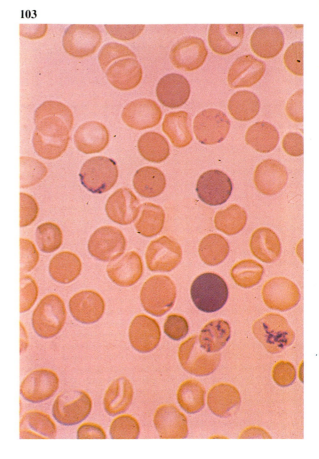

103

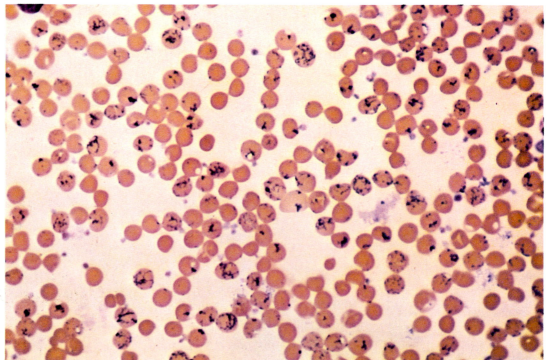

104

105. Anisocytosis. The red cells show variations in size. This smear illustrates minimal changes. Grosser anisocytosis is usually accompanied by other defects, such as poikilocytosis.

106. Poikilocytosis. The red cells show variations in shape. This smear also illustrates minimal changes, mostly of 'tear-drop' character – dacryocytosis. Grosser poikilocytosis usually occurs together with other defects.

107. Hypochromia and microcytosis – a general low-power view. A few cells stain normally and are of normal size (are normochromic and normocytic) but most have conspicuous central pallor (hypochromia) and small diameter (microcytosis).

106

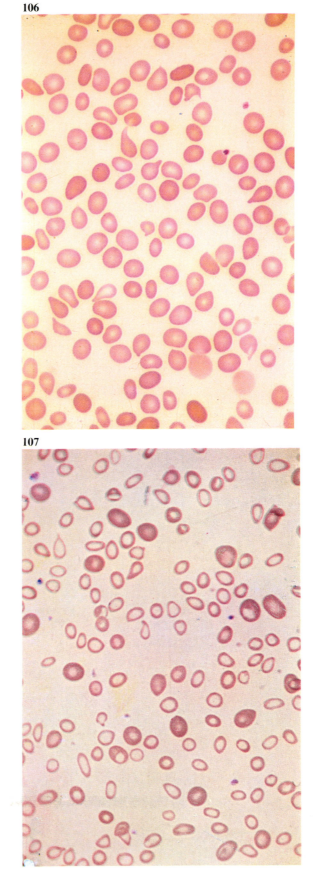

105

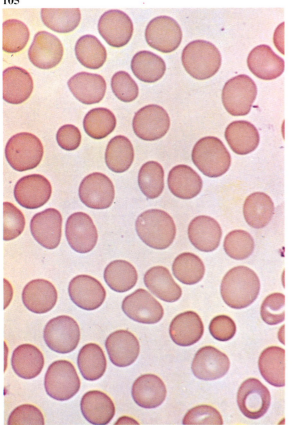

107

108. Hypochromia, anisocytosis and poikilocytosis; from a patient with iron deficiency anaemia.

109. Marked hypochromia, anisocytosis and poikilocytosis. Where, as here, irregular fragments of red cells or very grossly distorted cells are seen, the term 'schistocytosis' may be applied.

108

109

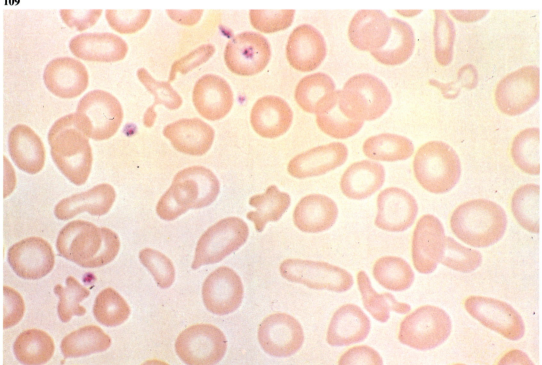

110 and 111. Respectively low- and high-power views of spherocytes. These are red cells with more spherical shape than normal, recognizable in stained smears by their apparent small diameter and dense staining.

  More than half the cells in these fields are acceptable as normal; only the small, very deeply stained cells can be regarded with fair certainty as spherocytes.

112. Anisocytosis due to a mixture of polychromatic reticulocytes and smaller spherocytes from a case of hereditary spherocytosis.

111

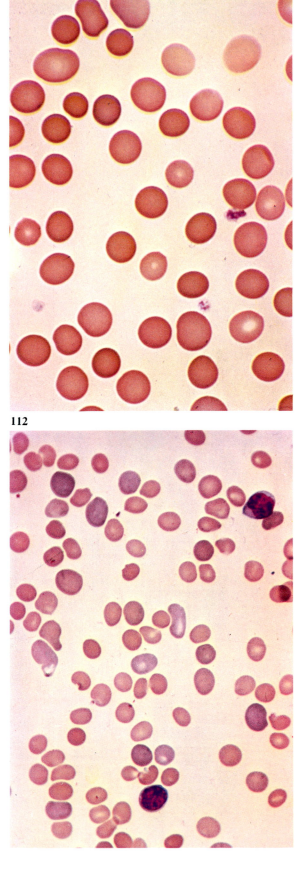

110

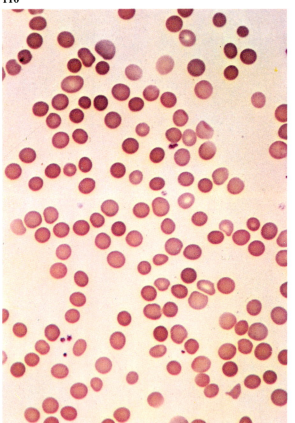

112

113. Elliptocytosis. An inherited anomaly of the red cells which involves, in this case, some 50 per cent of the red cell population. The affected cells appear oval or cigar-shaped.

114. Pseudo-elliptocytosis. A common artefact of smearing, recognizable as such by two chief points:
(a) the pseudo-elliptocytes are found only in certain areas of the smear, usually near the tail, and
(b) their long axes are generally parallel or nearly so, whereas true elliptocytes show random scattering of disposition (as in **113**).

115. 'Crescent cells' – disrupted erythrocytes drawn out into a crescentic form. This is a fairly common artefact of smearing, especially in anaemic blood.

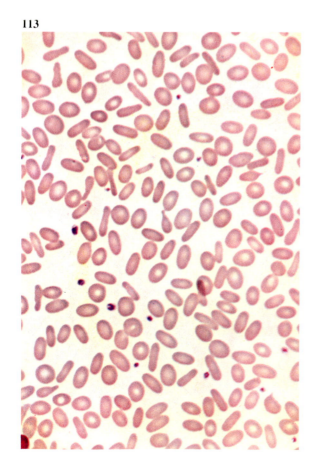

114

113

115

116. 'Crenation.' Crenated erythrocytes show an irregular undulation of the cell membrane in fixed smears. The appearance may be induced by exposing red cells to hypertonic saline or otherwise causing loss of fluid from the cells. In vivo this defect usually accompanies other defects of size, shape and structure. When the feature is marked, as here, the term 'burr cell' or echinocyte may be applied.

117. Erythrocytes showing a tendency to crenation and also with occasional constrictive 'cottage-loaf' defects in shape.

118. So-called 'spur cells' or acanthocytes showing some resemblance to crenated cells, but with sharper projections. They occur, often in association as here with fragmented schistocytes, especially in uraemia and microangiopathy, but may also be seen after splenectomy and in abetalipoproteinaemia and malabsorptive states.

117

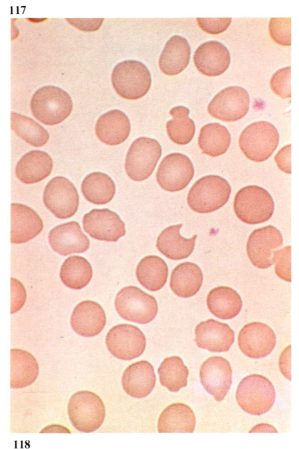

116

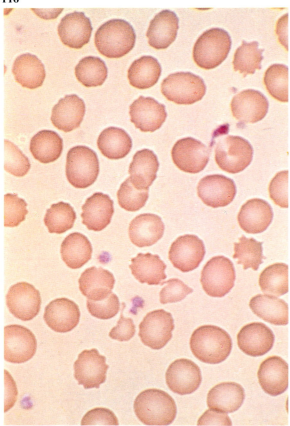

118

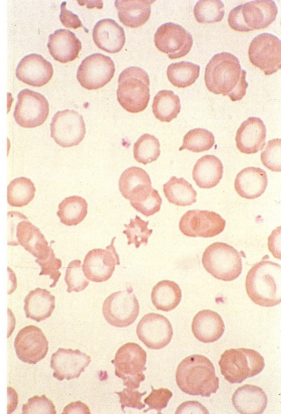

119. Moderate rouleaux formation; some hypochromia.

120. More marked rouleaux formation; this and the presence of a plasma cell suggests the possible diagnosis of myeloma.

121. Macrocytic erythrocytes from pernicious anaemia, with Howell-Jolly bodies in two of them.

120

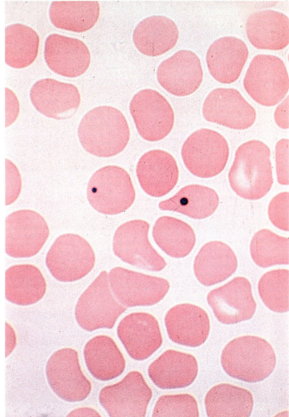

119

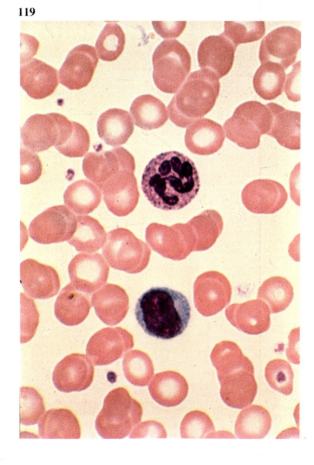

121

122. Target cells or codocytes, with accompanying aniso- and poikilocytosis and hypochromia from a case of ß-thalassaemia minor. The field also shows thin flat cells or leptocytes.

123. Another field from the same specimen, where target cells greatly predominate.

124. Poikilocytosis, with conspicuous sharp-angled 'helmet' cells, from a case of thrombotic thrombocytopenic purpura.

123

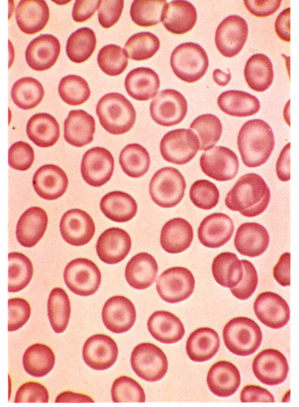

122

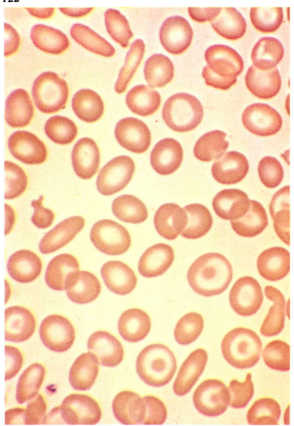

124

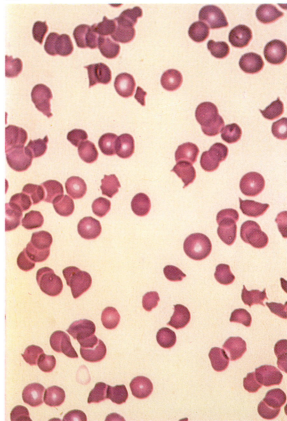

125. Target cells, macrocytes, spherocytes, schistocytes including helmet cells, and small cell fragments, from a case of severe ß-thalassaemia.

126. Leishman stained, fresh, fixed blood smear from a patient with sickle cell disease. Elongated or sickled cells are rare, but aniso- and poikilocytosis and target cells are more common.

127. Sickle cells, drepanocytes, formed by exposing erythrocytes from a patient with sickle cell disease to the reducing action of sodium metabisulphite under a sealed coverslip. As the reduced Hb S crystallises within the cells, they all come to assume the distorted, elongated, sickle shape.

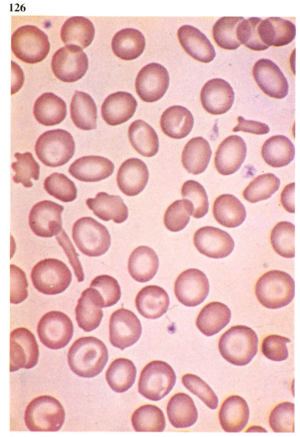

126

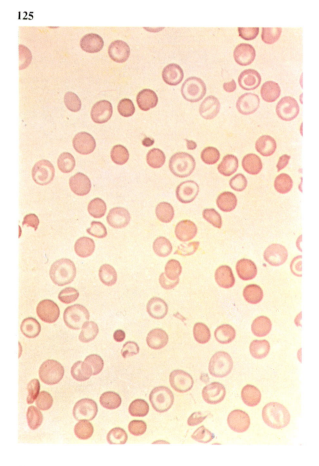

125

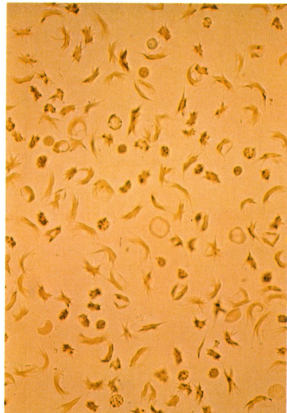

127

128 and 129. Cabot rings, respectively round and in figure of eight form, in stippled red cells. They are probably an RNA or protein precipitation artefact of little diagnostic significance but should not be confused with malaria parasites.

130. Leishman stained erythrocytes from a patient with haemolytic anaemia and increased iron stores. There is faint polychromasia present and also Howell-Jolly bodies. Granules of free iron are occasionally detectable.

131. Prussian blue stain for free iron on the same blood specimen as in **130**. Siderocytes (erythrocytes with free iron) are seen to be numerous.

130

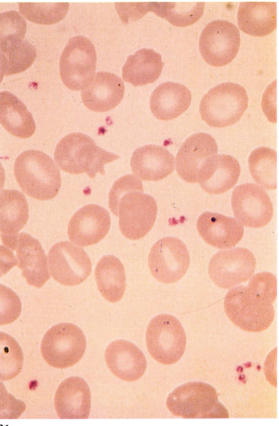

128

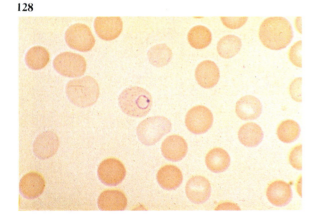

129

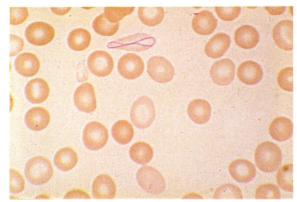

131

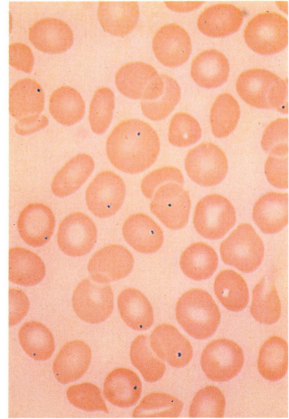

132. Heinz bodies in erythrocytes from haemolytic anaemia associated with deficiency of glucose-6-phosphate dehydrogenase.

133. Uncounterstained preparation of red cells from α-thalassaemia exposed to cresyl blue stain for 10 minutes – reticulocytes have taken up the stain.

134. A similar preparation from the same blood sample after 1 hour's exposure to stain. The multiple dotted positivity of 'golf-ball' pattern in many cells is that of haemoglobin H inclusions.

133

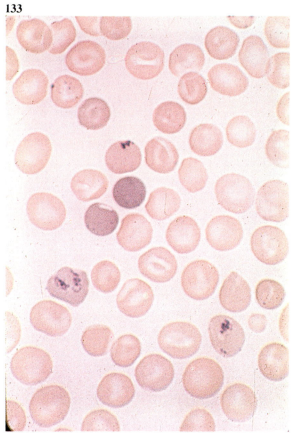

132

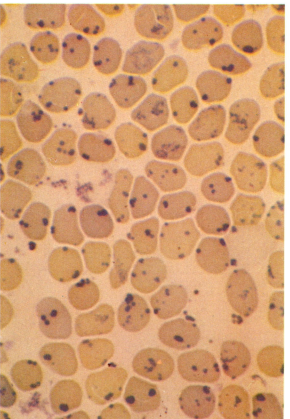

134

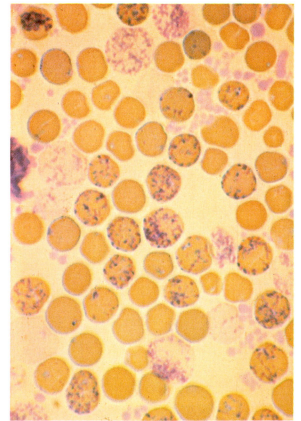

135 and 136. Target cells, spur cells or acanthocytes with multiple sharp projections, schistocytes of irregular fragmented shape and 'sputnik' cells, acanthocytes with two or three elongated spurs. Such changes are seen especially in microangiopathy and may be conspicuous in splenectomised patients.

137. Frequent conspicuous tear-drop poikilocytes – dacryocytes – in myelofibrosis.

136

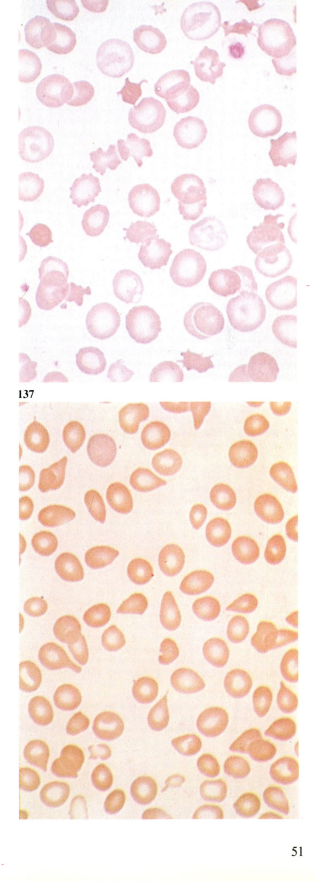

135

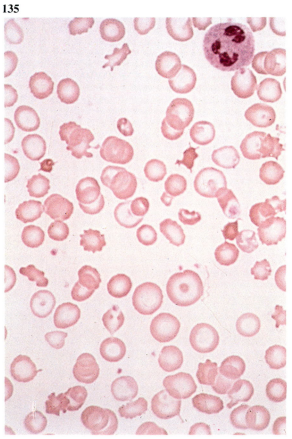

137

Part 2
Granulocytes, monocytes and megakaryocyte
Normal and abnormal forms

Although there probably exists a basic common progenitor or precursor cell for both myelopoiesis and lymphopoiesis, it is appropriate to group together the granulocyte, monocyte and the megakaryocyte-platelet series of cells because they share with each other and with the erythroblast series a less primitive functional stem cell in the bone marrow, and because they are frequently involved together in pathological processes.

The earliest recognizable cell of the granulocyte series is the myeloblast, which gives rise to a sequence of promyelocyte, myelocyte, metamyelocyte, stab cell and polymorph. From promyelocyte onwards 'specific' granules, neutrophil, eosinophil or basophil, become increasingly conspicuous in the cytoplasm and differentiate the common neutrophil granulocytes from the much less common eosinophils and the normally rare basophils. Mitoses may occur in immature cells up to the metamyelocyte stage, but there is now some evidence that the myelocyte may act as a secondary 'stem' cell for the granulocyte series, maintaining its own numbers and feeding cells into the maturation chain without normally requiring frequent replenishment from earlier precursors. In states of granulocyte hyperplasia and in granulocytic leukaemias this capacity is ineffective and earlier precursors derived from the primary stem cell preceding the myeloblast become more numerous.

Monocytes and their precursors – monoblasts and promonocytes – are present only in small numbers in normal marrow, and become conspicuous only in states of pathological proliferation or accumulation, notably leukaemias, and then nearly always in association with parallel involvement of the granulocyte line. The illustrations of monocyte precursors shown here are therefore taken from leukaemic states.

Myeloblasts and monoblasts may be difficult to differentiate in Romanowsky stained preparations, and may not always be easily distinguished from lymphoblasts or even from proerythroblasts in some leukaemic and erythraemic states. Cytochemical reactions will usually resolve such difficulties. Characteristic cytochemical patterns helpful in this regard are therefore included among the illustrations where appropriate.

Megakaryocytes usually present no difficulty in recognition, but their 'megakaryoblast' precursors are much smaller, and they too have similarities with other immature precursors of the myeloid series. Early but recognisable stages in the megakaryoblast-megakaryocyte line may be seen in idiopathic thrombocytopenic purpura and occasionally may be distinguished in acute myeloid leukaemia, but the earliest cells committed to this line, which are presumed by analogy to resemble the myeloblast and monoblast, cannot be certainly differentiated even by cytochemical means.

The illustrations have been chosen to allow the range of appearances shown by different cell types to be appreciated and to allow frequent comparisons between cells of different lines or of different stages of maturity. The particular importance of cytochemical reactions among the cells dealt with in this section is emphasized, especially in the differentiation of the varieties of acute myeloid leukaemia.

138. A myeloblast with two conspicuous nucleoli, a vacuole and a few azurophilic granules, a larger promyelocyte, a myelocyte and a stab cell of the neutrophil series together with a late normoblast.

139. A sequence of granulocytes, with a myeloblast, promyelocyte, myelocyte (with neutrophil granules), and a late neutrophil metamyelocyte or early stab cell.

140. A promyelocyte, two myelocytes, a metamyelocyte and a stab cell, of the neutrophil series.

139

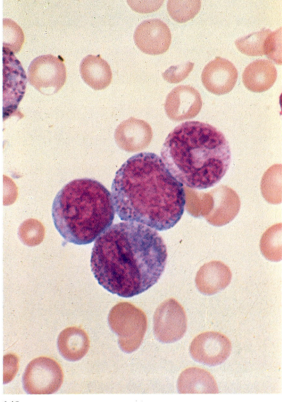

138

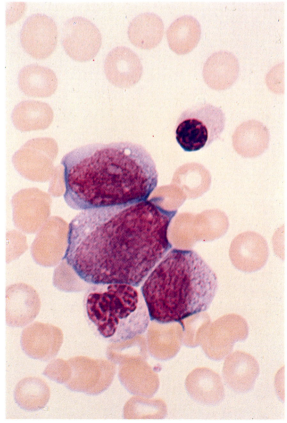

140

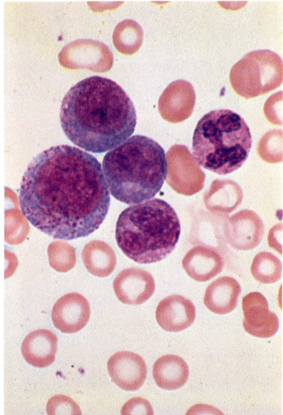

141. A myelocyte and three metamyelocytes of increasing maturity (with increasingly indented nuclei).

142. A myeloblast and an early neutrophil myelocyte.

143. A myeloblast with a monocyte for comparison. The monocyte shows a coarser nuclear chromatin pattern, absence of nucleoli, and a greyish rather than basophilic cytoplasm.

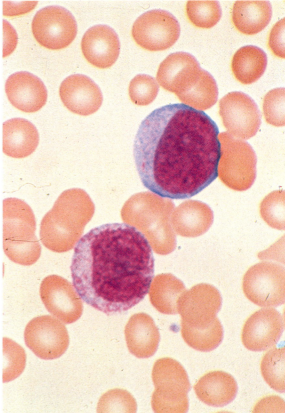

142

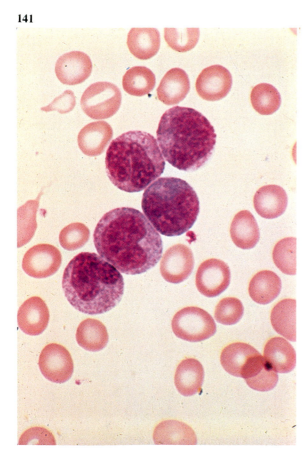

141

143

144. A myeloblast with a neutrophil stab cell.

145. A myeloblast or early promyelocyte with a neutrophil segmented polymorph.

146. Two myeloblasts, the smaller showing diminution of cytoplasmic basophilia and the first appearance of azurophilic granulation, a promyelocyte with coarse granularity and considerably larger size than the myeloblasts (a frequent finding), and a monocyte, for comparison.

145

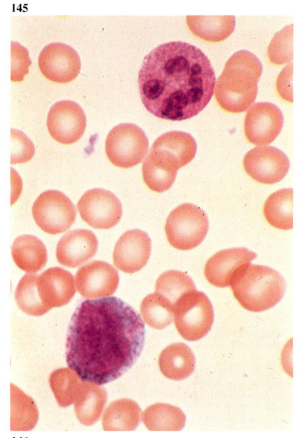

144

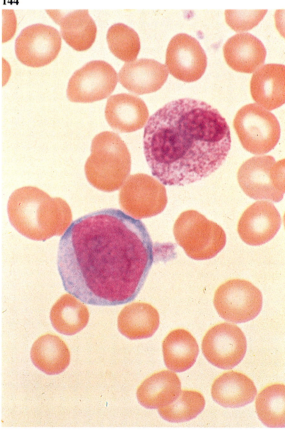

146

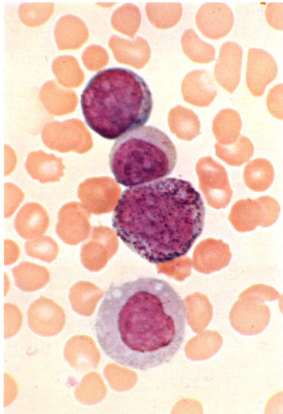

147. A sequence of neutrophil granulocytes with a myeloblast, promyelocyte (with vacuole), myelocyte, metamyelocyte and two stab cells.

148. Peroxidase reaction on normal bone marrow cells; a sequence of granulocyte precursors shows strong positivity; erythroblasts are negative.

149. Sudan black B (SB) stain on normal marrow, illustrating the increasingly heavy positivity in the developing granulocytes. Erythroblasts are negative as is a lymphocyte at the bottom right.

148

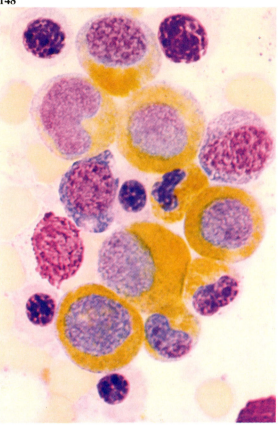

147

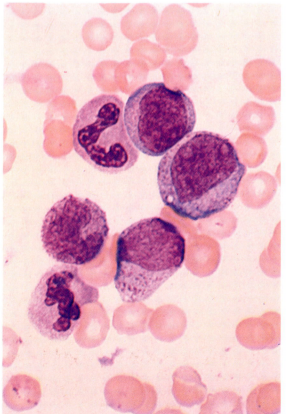

149

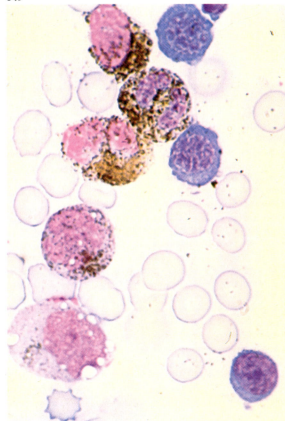

150. SB stain on normal buffy coat smear. The coarse positivity in neutrophil polymorphs contrasts with the weaker reactions in monocytes – one negative, one with a few fine granules and three with scattered granules, more discrete than in the neutrophils. There is an eosinophil with hollow positive granules, a negative basophil and a negative lymphocyte.

151. SB stain on normal peripheral blood, illustrating the contrast between the dense and coarse positivity in mature granulocytes and the discrete scattered granule pattern of monocytes. Lymphocytes are negative.

152. SB stain on normal buffy coat showing the presence of a myeloblast, as can usually be found in normal circulating blood if carefully sought. Two negative basophils, two normally reacting neutrophil polymorphs, a monocyte and a lymphocyte complete the field.

151

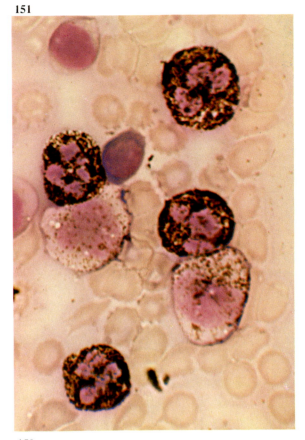

150

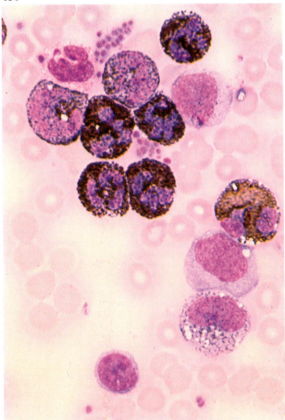

152

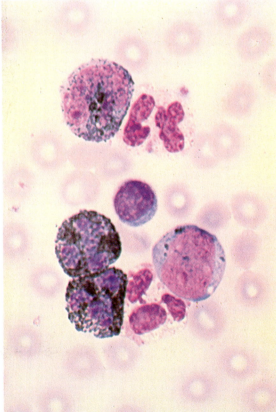

153. Peroxidase stain on normal peripheral blood. Two neutrophil polymorphs are strongly positive; a basophil is negative and one of two monocytes shows localised cytoplasmic positivity contrasting with the discrete scattered granules of the SB reaction in monocytes.

154. PAS reaction on a normal bone marrow smear, to illustrate the gradual increase in positivity found in the granulocyte series with increasing cell maturity. Erythroblasts are negative. The disrupted cell with centrally situated nucleus and coarse granules of material containing free iron, spreading out between neighbouring cells, is a reticulo-endothelial cell.

155. PAS reaction on normal buffy coat showing two positive neutrophils, a coarsely reacting basophil, two negative lymphocytes, two monocytes with faint diffuse reaction and an eosinophil with negative granules against a PAS positive background.

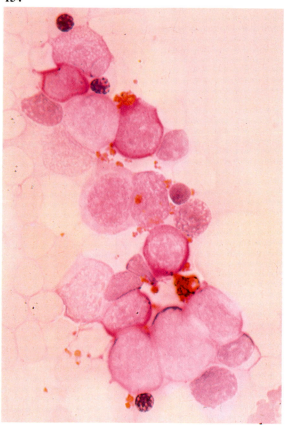

154

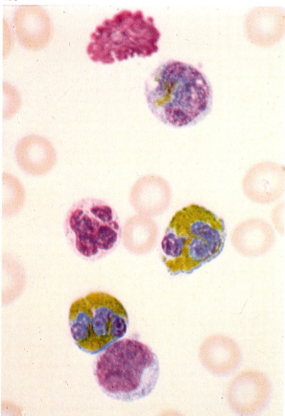

153

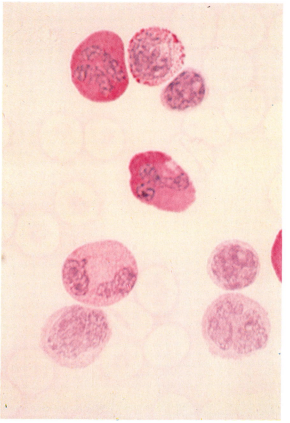

155

156. Acid phosphatase in granulocytes in normal marrow; myeloblast, promyelocyte, stab and two segmented neutrophils.

157. Dual esterase, on normal marrow cells – 7 neutrophil granulocytes including two myelocytes, one in mitosis, all showing chloroacetate esterase (CE) positivity, two negative eosinophil myelocytes, a monocyte of mixed butyrate esterase (BE) and CE reaction and a negative late normoblast and lymphocyte.

158. Dual esterase on normal peripheral blood showing CE positive polymorph and BE positive monocyte.

157

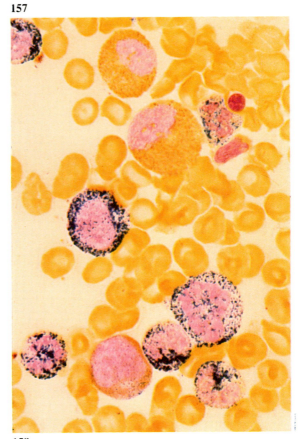

156

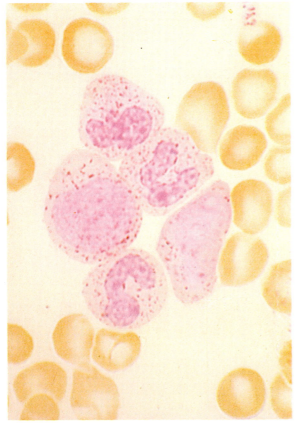

158

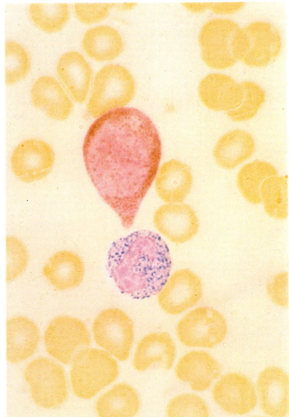

159–164. *Phases of mitosis in blast cells from acute myeloid leukaemia.*

159 and 160. Early and more advanced prophase.

161. Late prophase.

162. Metaphase.

163. Metaphase progressing to anaphase.

164. Anaphase.

161

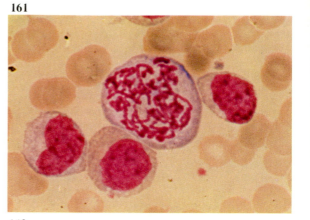

162

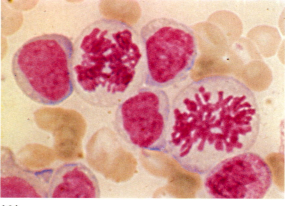

159

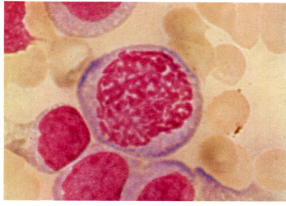

160

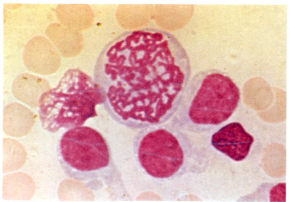

163

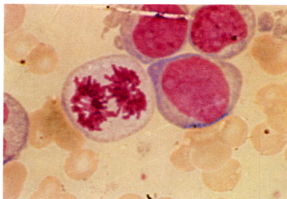

164

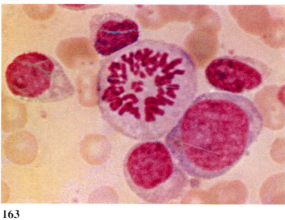

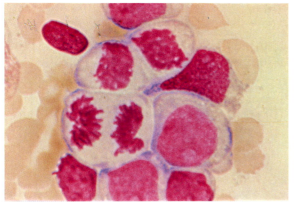

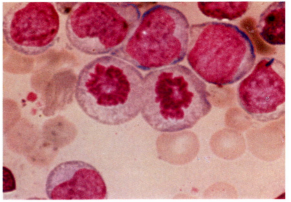

165 and 166. *Further phases of mitosis in blast cells from acute myeloid leukaemia.*

165. Late anaphase.

166. Telophase.

167. G-banded normal metaphase spread, with paired karyotype.

167

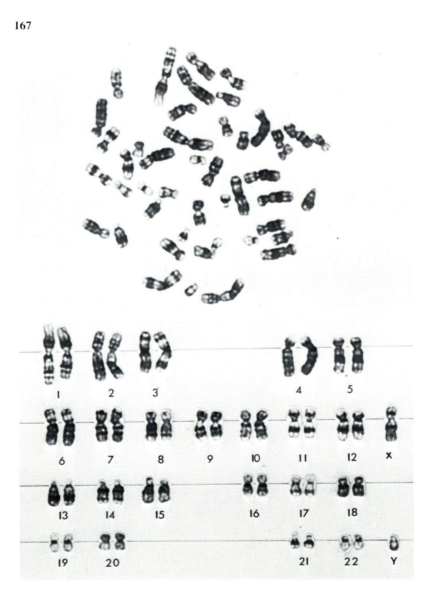

168. A group of myeloblasts from an acute myeloid leukaemia (AML) showing Auer rods and azurophilic inclusions. There is also a myelocyte and two lymphocytes in this field.

169. Peroxidase reaction in AML, showing positivity virtually confined to Auer rods which are present in all the blast cells in this field. A single lymphocyte is negative.

170. Myeloblasts and polymorphs from a similar case of AML. Several myeloblasts show peroxidase positivity, and inclusions appear positively stained. The polymorphs in this case show strongly positive reaction as in normal polymorphs – although sometimes in AML the mature polymorphs show weak or even negative reactions.

171. Sudan black reaction in AML, showing several strongly positive Auer rods and localised reaction.

172. Sudan black reaction in myeloblasts of AML. Strong positivity is mostly localised to cytoplasm with a positive Auer rod in one myeloblast.

169
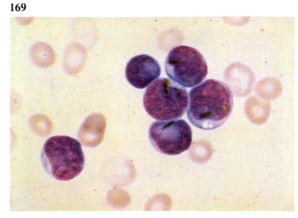

170
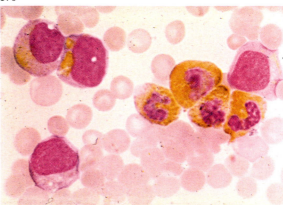

171
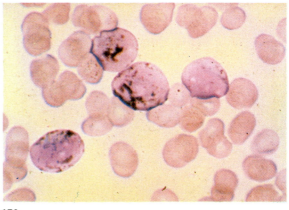

172
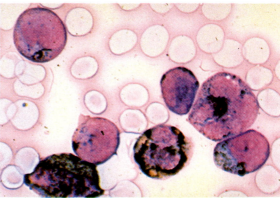

168
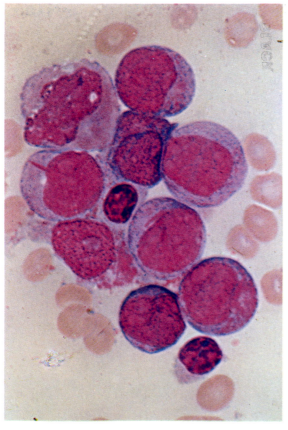

173. PAS reaction in a similar acute leukaemia. The single polymorph is normally positive, while the primitive cells give a reaction ranging from negative to a weak diffuse positive tinge over most of the cytoplasm. Sometimes fine granules are also present, but on a background of diffuse tingeing unlike the clear background seen in lymphoblasts.

174. Acid phosphatase reaction in a similar case. Moderately coarse granular positivity is present.

175. Dual esterase reaction in the same case. The myeloblasts show scattered CE positivity. A single neutrophil polymorph shows strong CE reaction.

174

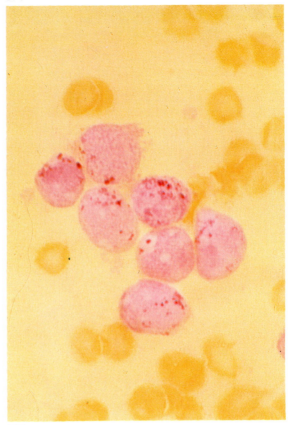

173

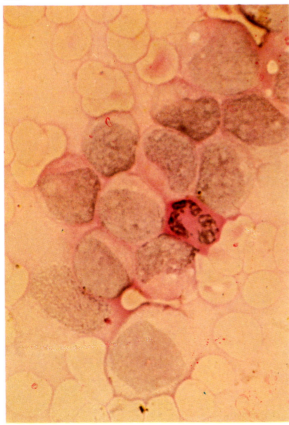

175

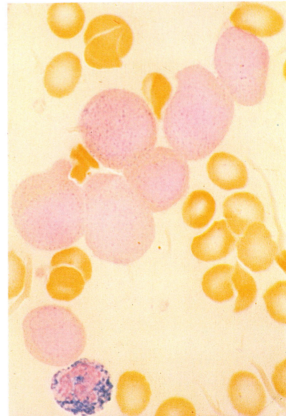

176. An AML with early granularity in the archoplasmic zone next to the nucleus in most cells. The cells are leukaemic promyelocytes, but do not show multiple Auer rods or conspicuous nuclear changes of the type often seen in acute promyelocytic leukaemia (APL). Striking vacuolation was a feature of this case.

177. Sudan black reaction in the same case. The dense cytoplasmic positivity resembles that seen in normal later granulocytes from myelocyte to polymorph.

178. The PAS reaction on these cells shows diffuse positivity with some increased granularity in the archoplasmic zones. This is the PAS picture of acute myeloid leukaemia with promyelocytic preponderance.

179. Dual esterase reaction in the same cells, with strong CE positivity, including a CE positive Auer rod.

177

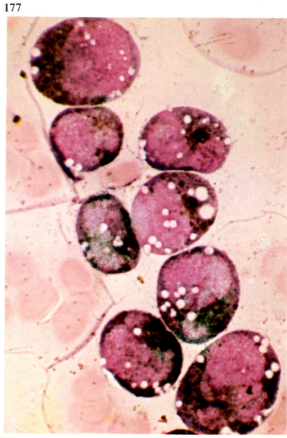

176

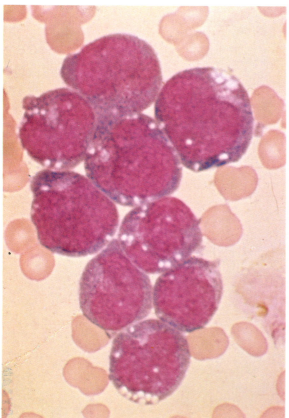

178

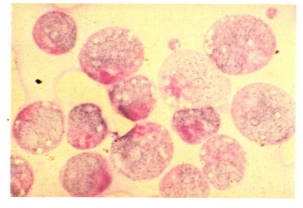

179

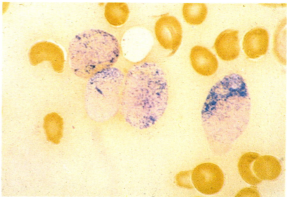

180. Acute promyelocytic leukaemia (APL) with coarse azurophil (primary) granules largely filling the cytoplasm of all the leukaemic promyelocytes. Nucleoli are conspicuous and variable in number but the overlapping or dumb-bell type of nuclear shape often seen in APL is not shown here, and Auer rods are not visible in this Leishman-stained preparation.

181. Another preparation from the same case, stained with MGG. Conspicuous multiple Auer rods are now visible in several cells. The difference in staining reaction of Auer rods in APL to Leishman and MGG stains, though not always shown, is frequently striking.

182. Another case of APL which showed few or no cytoplasmic granules and no Auer rods with either Leishman or MGG stains, but with the characteristic and diagnostic nuclear shape with dumb-bell appearance or overlapping of twinned nuclear lobes.

181

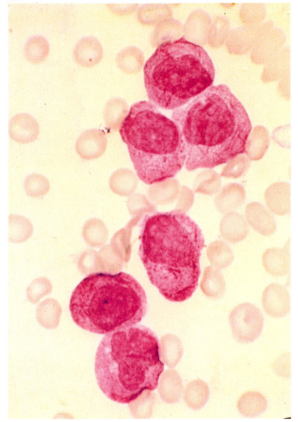

180

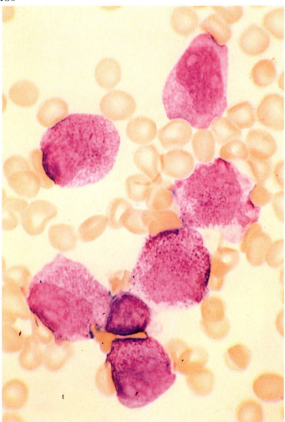

182

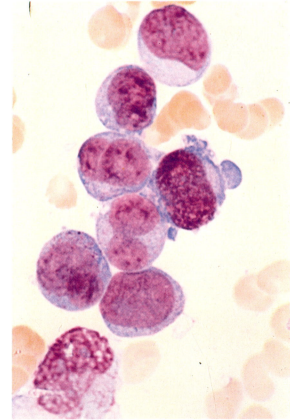

183. The same marrow after 20 hours in culture; coarse azurophil granules have now appeared in the cytoplasm of all the promyelocytes.

184. Marrow from APL: SB stain showing heavy overall positivity, with positively stained Auer rods.

185. Another case of APL: SB stain of marrow cells shows similar overall positivity, but here the Auer rods are SB-negative.

184

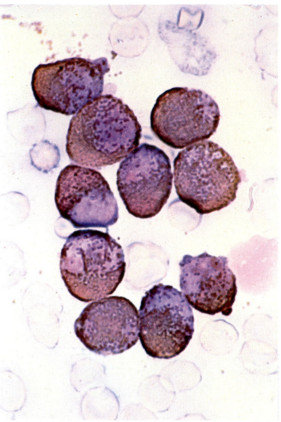

183

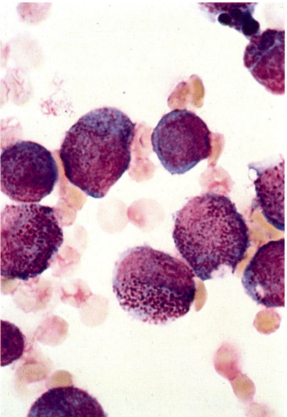

185

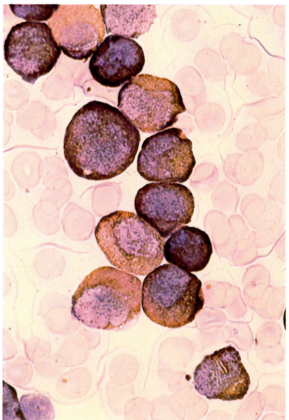

186. PAS reaction in APL, showing moderately strong diffuse and finely granular positivity. Occasional nuclei show the APL-type of twinning or distortion.

187. Another field from the same preparation, showing unusually strongly PAS-positive Auer rods.

188. Acid phosphatase reaction in APL; the leukaemic promyelocytes show only very weak granular reaction, contrasting with the stronger reaction in a neighbouring plasma cell.

187

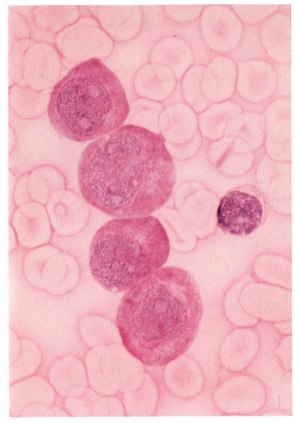

186

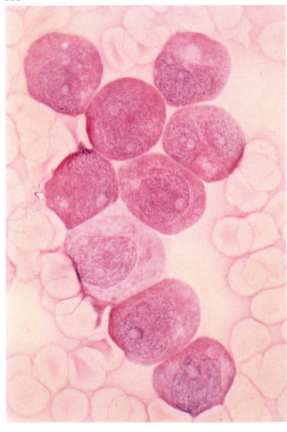

188

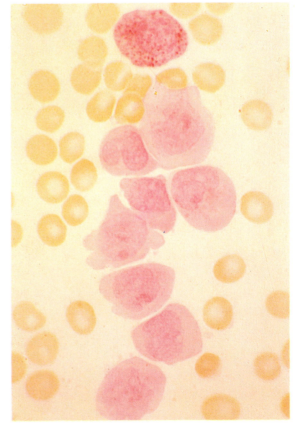

189. Dual esterase reaction in APL: a negative early erythroblast and two CE positive promyelocytes containing multiple Auer rods showing CE positive outlines but negative cores.

190. Dual esterase reaction in APL: the leukaemic cells show CE positivity with numerous strongly CE positive Auer rods; the dense circular or ring-like CE-positive structures, some with a hollow CE-negative centre, shown especially in one of the promyelocytes, probably represent early stages in the formation of Auer rods. A sequence between these hollow rings and the more common hollow rods (as shown in **189**) can be made out.

191. Dual esterase reaction in another case of APL: the hollow rings and short Auer rods in several of the promyelocytes here display a CE-positive envelope with a BE-positive core.

190

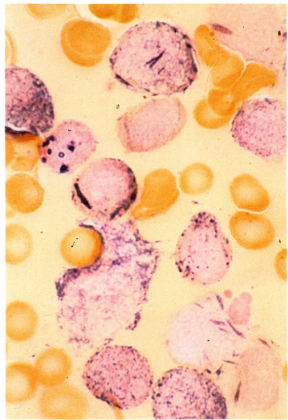

189

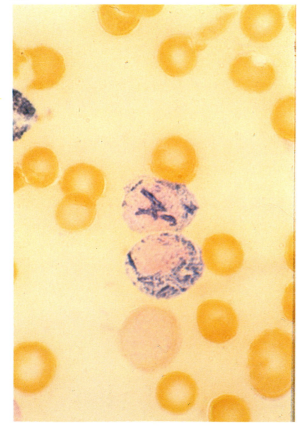

191

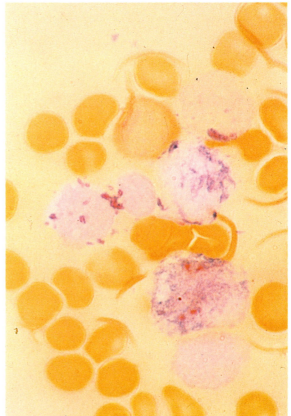

192–196. *A further APL variant.*

192. APL variant – the cells show numerous moderately coarse granules but no Auer rods were visible in either Leishman (here) or MGG stains.

193. Heavy granular SB positivity typical of promyelocytes but again no Auer rods.

194. Strong diffuse tinge of PAS positivity – the characteristic APL pattern.

195. Negative reaction or fine scattered granules only of acid phosphatase positivity.

196. Dual esterase – a remarkable mixture of strong reactions to both CE and BE perhaps shared sometimes in the same granule (cf. **191**).

193

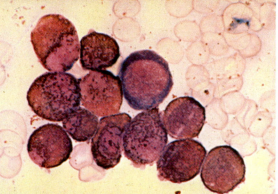

194

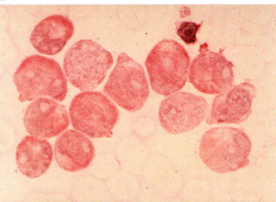

192

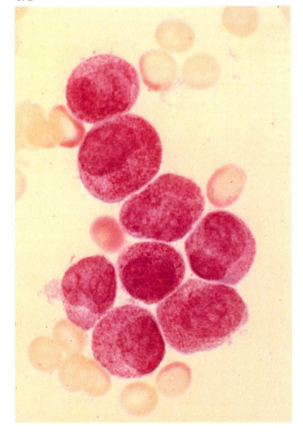

195

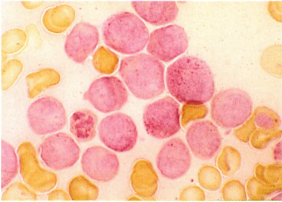

196

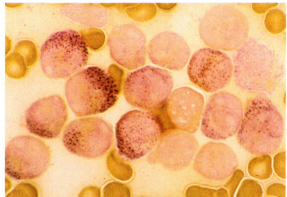

197. Exceptionally large promyelocytes, myelocytes and metamyelocytes may be encountered in the marrow of pernicious anaemia and sometimes of other megaloblastic (and rarely normoblastic) anaemias. Here are examples of giant promyelocyte and myelocyte from P.A.

198. A neutrophil stab cell and a segmented cell from normal peripheral blood.

199. Three neutrophil segmented polymorphs, one showing a drumstick appendage. Two small sessile appendages, not countable as drumsticks, are visible in the upper cell.

198

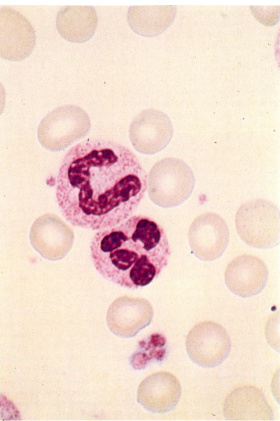

197

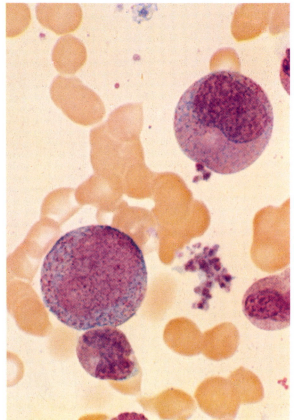

199

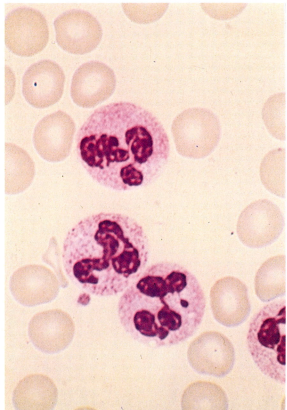

200. Three segmented neutrophil polymorphs and a myelocyte.

201. A drumstick appendage attached to the nucleus of a stab cell.

202. A segmented neutrophil polymorph, with drumstick appendage and a monocyte.

201

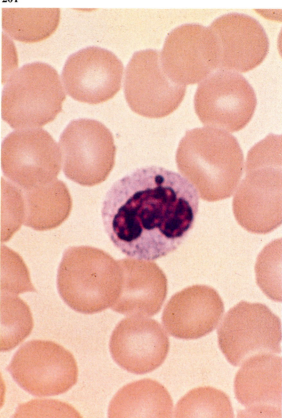

200

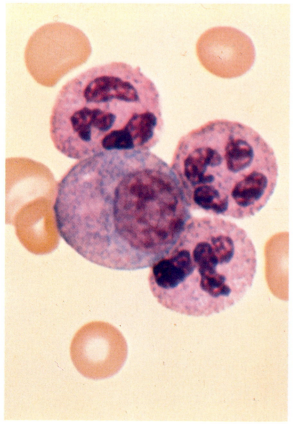

202

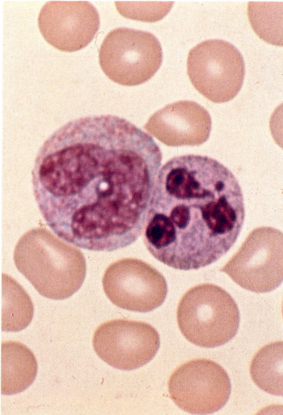

203. Three segmented neutrophil polymorphs in the centre of the field, with an eosinophil polymorph and a monocyte. Two more neutrophils appear at the edges of the field. How many drumstick appendages are present?

204. Nuclear twinning in neutrophil polymorphs; there appears to be something approaching a mirror image disposition of nuclear lobes. This specimen was from an acute myeloid leukaemia following chemotherapy.

205. A multi-lobed polymorph from the peripheral blood in a megaloblastic anaemia.

206. Coarse granularity in stab cells from an infective state with leucocytosis and left shift. This appearance is sometimes called toxic granularity.

204
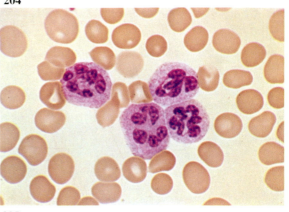
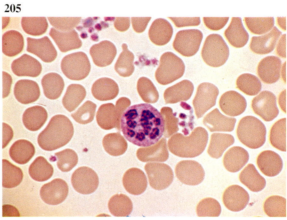

205
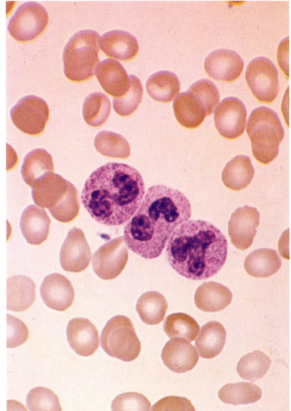

203
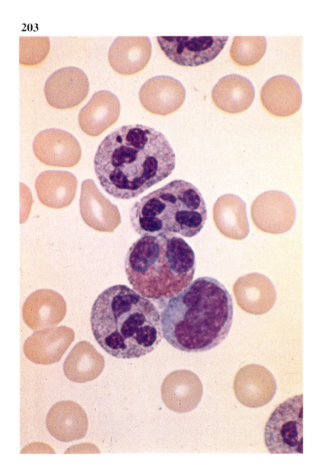

206

207. A normally granular and non-granular neutrophil polymorph, together with a lymphocyte. Absence of granularity in polymorphs is most common in leukaemic states but may occasionally be seen in most anaemic or leucopenic states.

208. A poorly granular, multi-lobed polymorph from the bone marrow in a case of pernicious anaemia. A second segmented cell with four nuclear lobes, a basophil polymorph and various earlier granulocytes make up the remainder of the field.

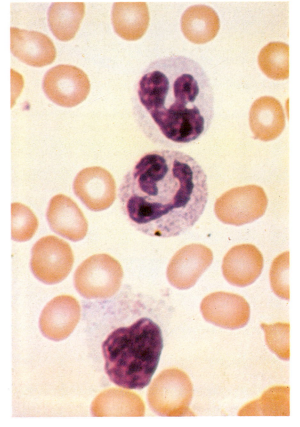

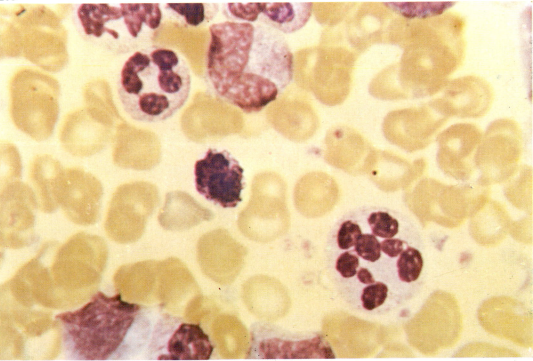

209. Leucocyte alkaline phosphatase (LAP) reaction. Weak and stronger positivity in neutrophils, with a negative lymphocyte. Positive reactions in haemic cells are virtually confined to neutrophil stabs and segmented cells, although macrophages may also show positivity.

210. Grades of positivity in polymorphs, ranging from 1 (+) to 4 (++++). Summation of ratings on 100 neutrophils gives a score with possible range from 0 to 400. The normal range is between 15 and 100.

211. Increased LAP score, with many strongly positive cells from an inflammatory leucocytosis.

Similarly high scores may be found in polycythaemia vera, myelofibrosis and Hodgkin's disease.

210

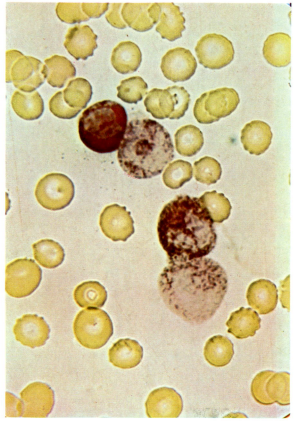

209

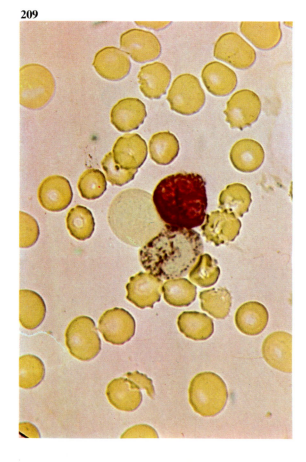

211

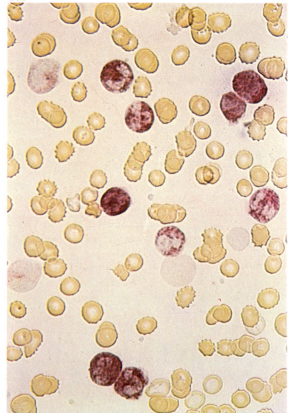

212. Eosinophil myelocytes and metamyelocyte in bone marrow. There is also a neutrophil myelocyte, various stab and segmented neutrophils and a late normoblast.

213. A normal eosinophil polymorph, showing the characteristic 'spectacle' arrangement of the nuclear segments, which are usually two in number.

214. A disrupted eosinophil.

215 and 216. Eosinophil polymorphs with agranular spaces in the cytoplasm. This appearance is not uncommon, but becomes most frequent and conspicuous in leukaemias.

213

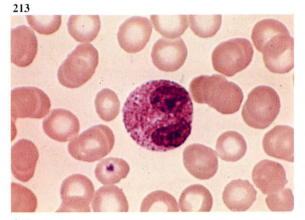

214

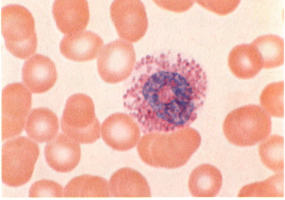

212

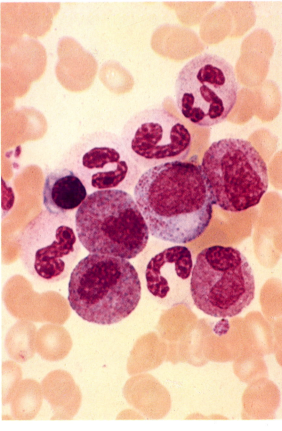

215

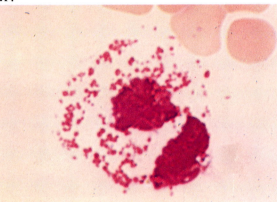

216

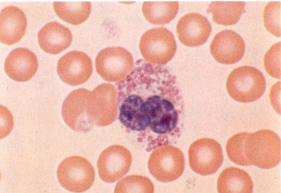

217. Familial eosinophilia; general view of bone marrow; all stages of maturation in the eosinophil series are present.

218. The same case: high power view to show 'amphophil' appearances (both basophil and eosinophil-staining granules) in eosinophil promyelocyte and myelocytes. The 'basophil' granules probably represent primary eosinophil granules.

219. Peripheral blood in the same case. Numerous mature eosinophil polymorphs are present.

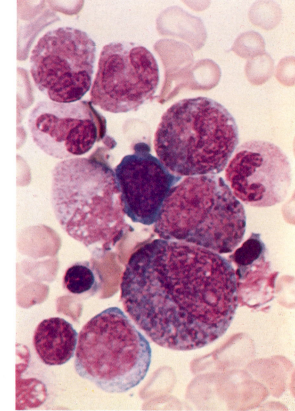

218

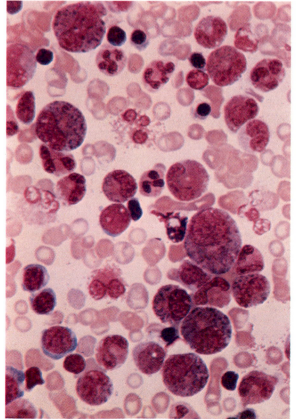

217

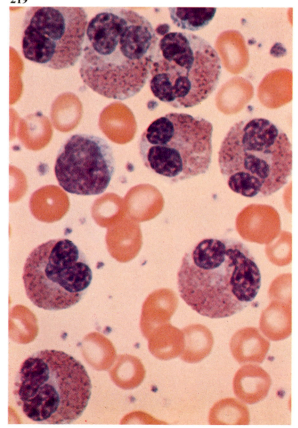

219

220. Basophil precursors, together with neutrophil precursors in the blood of a patient with chronic myeloid leukaemia.

221. A basophil polymorph, together with two neutrophils, an eosinophil and a lymphocyte.

222. Another basophil polymorph. These cells do not show the clear separation of nuclear lobes seen in mature polymorphs of other kinds but overlapping lobes may be distinguished.

223. A disrupted degenerating basophil, with loss of nuclear structure and scattering of granules.

221

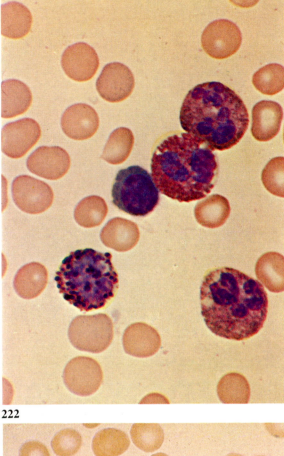

220

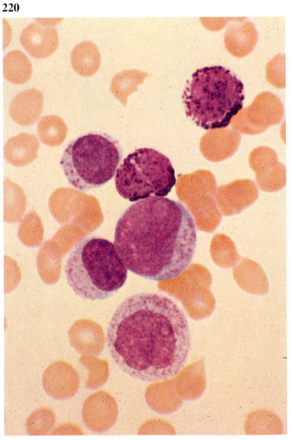

222

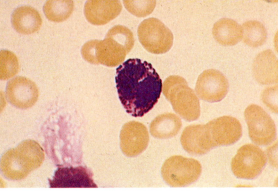

223

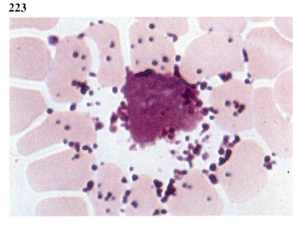

224. PAS reaction in mature polymorphs of neutrophil, eosinophil and basophil series. The central neutrophil has dense, strong, granular positivity, packing the cytoplasm to leave no visible background. The eosinophil shows granules with unstained, negative appearance against background positivity.

The basophil granules are discretely scattered, very heavily positive, against a negative or weakly stained background. Salivary amylase removes most positivity, but not that of the basophil granules, which presumably do not contain glycogen. The unstained granules in the eosinophil are certainly the specific granules seen in Romanowsky stains, but the PAS positive granules in the basophil are not identical with the specific granules in that cell.

225. SB reaction in neutrophil, eosinophil and basophil polymorphs. Neutrophil granules are positive, eosinophil granules show positivity with a hollow centre and basophil granules are negative (as here) or occasionally show positive or metachromatic staining.

226. Peroxidase reaction in a similar trio, with appearances generally similar to those in the SB reaction.

225
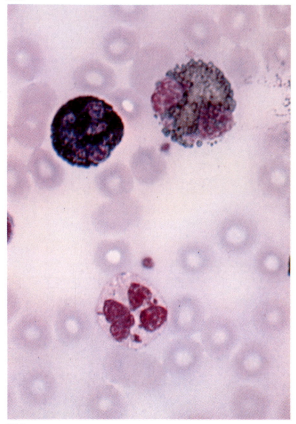

224
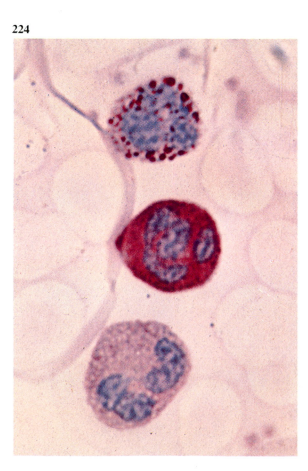

226
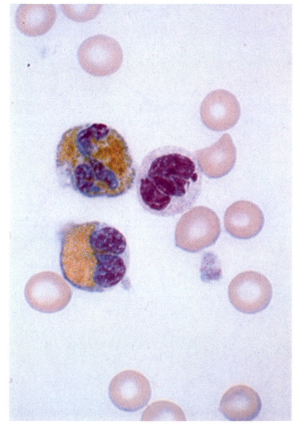

227. Alkaline phosphatase reaction in a similar cell group; only the neutrophil shows positivity.

228. Acid phosphatase reaction in neutrophil, eosinophil and basophil polymorphs and a monocyte. The neutrophil and monocyte show normally positive granular reactions; the eosinophil and basophil show very little positivity with only a rare positively-reacting granule.

229. Dual esterase in a group of basophil, eosinophil, two neutrophils and a monocyte. The basophil and eosinophil cells are essentially negative, while the neutrophils show typical CE positivity and the monocyte typical BE positivity.

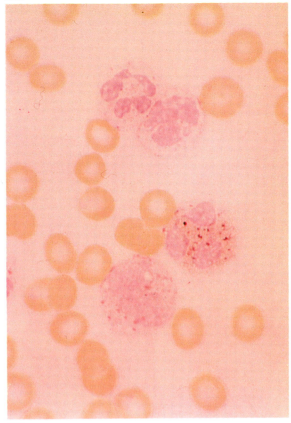

228

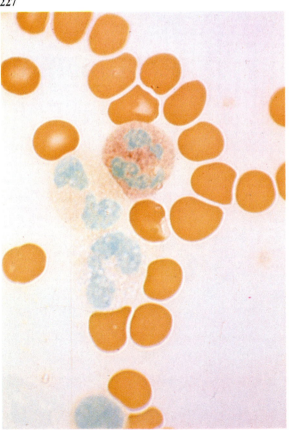

227

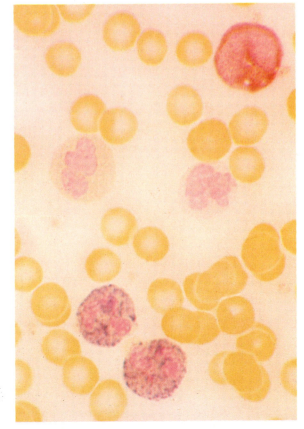

229

230. A low power view of peripheral blood from a patient with chronic myeloid leukaemia (CML). Granulocytes of all stages of maturation can be seen, mostly neutrophils but with occasional basophils.

231. A higher power view of another preparation showing various granulocyte precursors including several basophils. In addition to the granulocytes two normoblasts are present.

232. A chromosome spread from a male patient with CML. The Philadelphia (Ph¹) chromosome, seen at 12 o'clock, results from a translocation between the long arms of chromosomes 22 and 9 (usually) visible in unbanded preparations as a loss of material from the long arms of one of the small acrocentric chromosomes (no. 22).

231

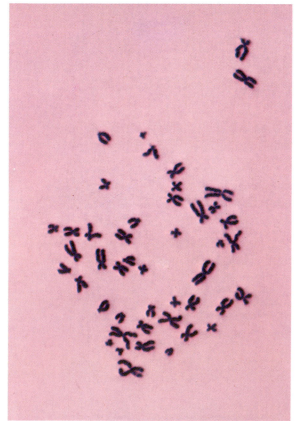

230

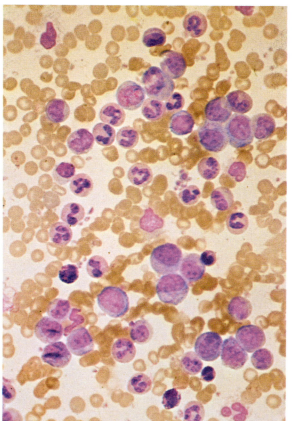

232

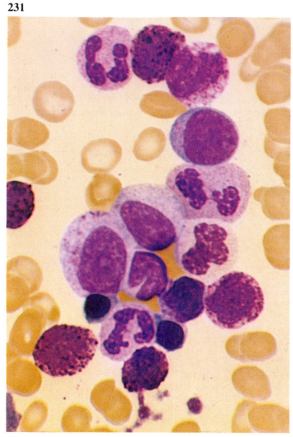

233. LAP reaction in CML. The polymorphs are virtually devoid of enzyme. This is a striking and almost uniform characteristic of neutrophils in CML.

234. Under effective treatment, when the blood picture returns quantitatively to normal, the LAP score remains low (and precursors still have the Ph¹ chromosome), but an improvement towards the lower normal range often occurs.

This preparation shows the presence of faint positivity in some polymorphs from a buffy coat in well-controlled CML.

235. When blastic crisis emerges, the LAP score usually rises sharply. This figure shows + to ++ reactions in polymorphs in a peripheral blood film from a patient at this stage of the disease. Surrounding the group of polymorphs are predominantly blast cells.

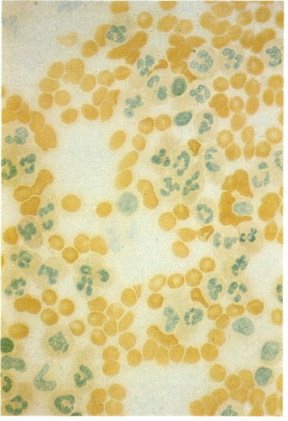

234

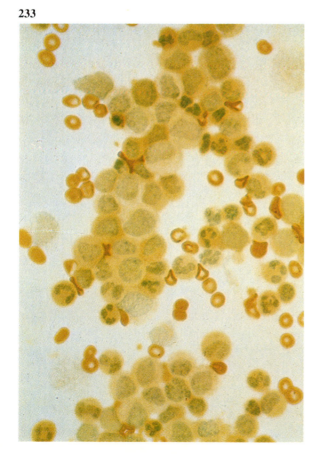

233

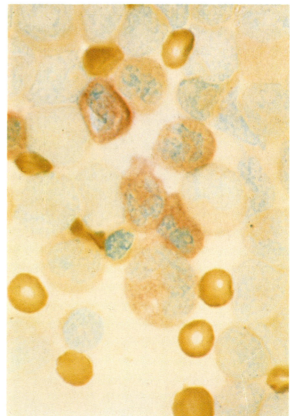

235

236. Peroxidase reaction in CML. Exceptionally strong positivity is the rule in polymorphs in this disease.

237. Sudan black reaction in CML. The polymorphs may react normally with strong positivity, as here, but the reaction is sometimes weaker than normal and negative polymorphs may be found, probably mostly basophils rather than neutrophils (cf. **239**).

238. Basophils may show a metachromatic reddish staining as in this field from CML progressing towards myelofibrosis, which also shows typical positivity in eosinophils and neutrophils with SB.

239. A field from CML showing SB positivity in neutrophil precursors, while 4 basophils show varying degrees of reaction, partly metachromatic. One is almost SB negative.

237

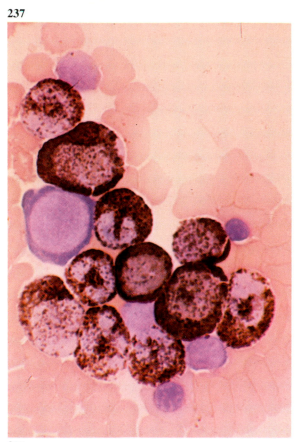

236

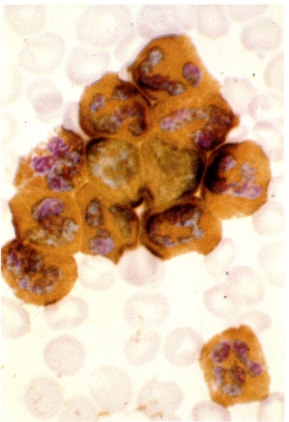

238

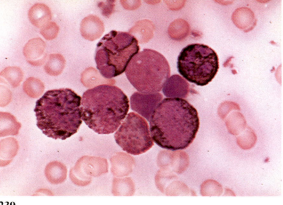

239

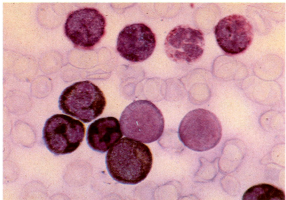

240–243. *CML progression to myelofibrosis/megakaryocytic myeloproliferative state.*

240. Romanowsky stain shows two blast cells, to left and right, several megakaryocyte nuclei and masses of variably sized platelets.

241. PAS reaction shows the coarse positivity in platelet and megakaryocyte precursors.

242. Acid phosphatase is strongly positive in megakaryocyte precursors.

243. Dual esterase shows strong CE reaction in polymorphs but only weak BE reaction in a megakaryocyte and some platelets and in a blast cell.

241

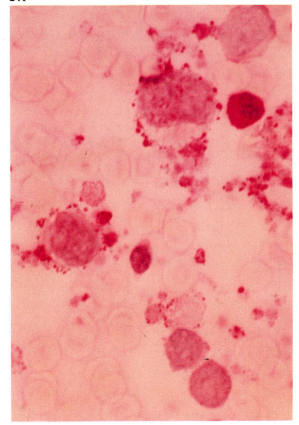

240

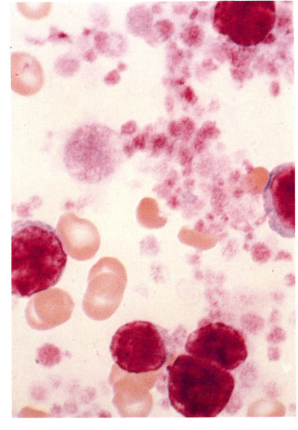

242

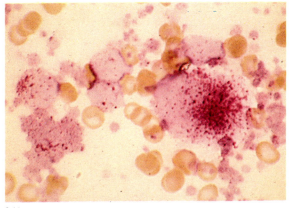

243

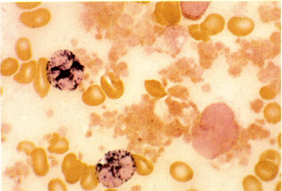

244–248. *CML progression to myeloblast-type crisis.*

244. Romanowsky stain shows a group of blast cells with little evidence of differentiation.

245. The SB stain shows that all contain localised strong cytoplasmic positivity of granulocytic type.

246. The blast cells are largely PAS negative or weakly reacting with diffuse positivity and fine granules. A monocyte/macrophage shows some coarse PAS positive granules – a polymorph is normally positive.

247. The acid phosphatase reaction shows a few coarse granules of positivity in most blast cells and a trail of strongly positive granules from a partially disrupting macrophage.

248. The dual esterase reaction displays moderately strong CE positivity in most of the blast cells.

245

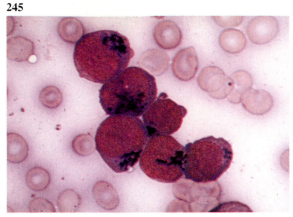

246

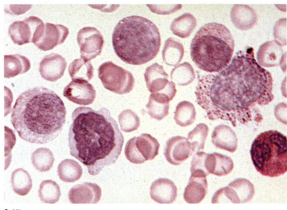

247

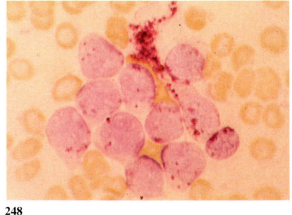

248

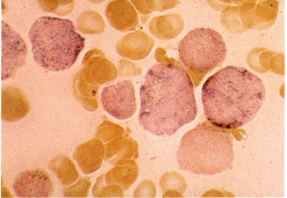

244

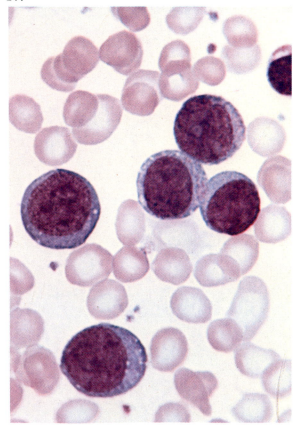

249–253. *CML progression to lymphoblast-type crisis, with cALL antigenic findings.*

249. Romanowsky stain shows agranular blast cells with high nuclear-cytoplasmic ratio.

250. The blast cells are negative to SB.

251. The PAS stain shows occasional block or coarse granular positivity of lymphoblastic type.

252. Acid phosphatase staining shows scattered positive reaction, not especially concentrated at one pole of the nucleus.

253. Dual esterase staining shows a strongly CE-positive granulocyte but only weak scattered BE positivity in the blast cells.

250
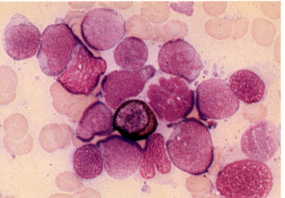

251
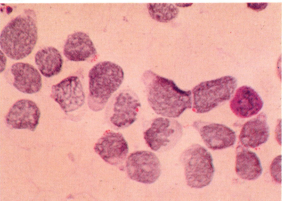

249
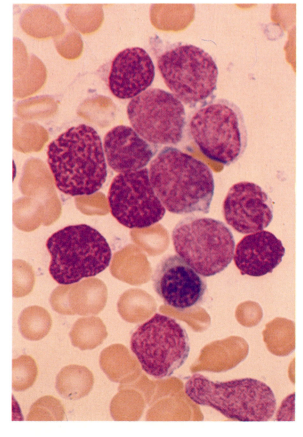

252
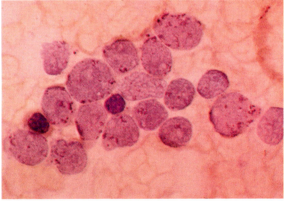

253
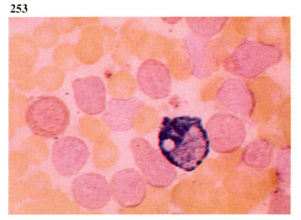

254–256. *Basophil predominance in malignant progression of CML.*

254. Romanowsky stain shows a mixture of granulocytes and their precursors including two blast cells, myelocytes, stab cells and four basophils.

255. SB stain in the same case, showing a myeloblast, three neutrophil myelocytes with normally strong sudanophilia and four basophils from promyelocyte to polymorph with mostly reddish metachromatic staining of the basophil granules.

256. PAS reaction in the same case, showing coarse and patchy positivity in four basophils at various stages of maturity from promyelocyte onwards. Neutrophil precursors show normal diffuse tinge.

255

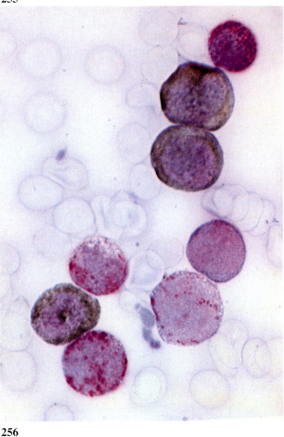

254

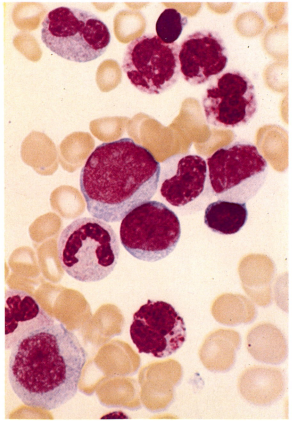

256

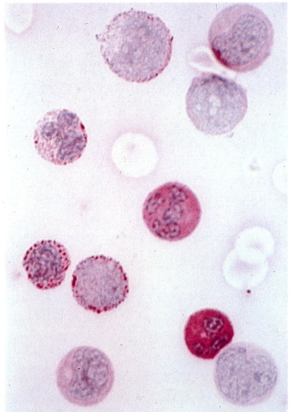

257–262. *Examples of defective granulogenesis – with a diffuse eosinophilic staining of the cytoplasm of myelocytes and few specific granules. This appearance is most often observed in preleukaemic states, including defective erythropoiesis with excess of myeloblasts, neutrophilic myelogranular dysplasia and smouldering leukaemia.*

257. Typical appearance of myelocytic cytoplasm in this condition as seen by Romanowsky staining.

258. A myeloblast, two probable premitotic promyelocytes with scattered azurophil granules and prominent nucleoli, five myelocytes with diffuse eosinophilic staining and two later granulocytes with a more normally granular cytoplasm.

259. Further granulocytes from the same case, showing the diffuse and finely granular eosinophilic staining in several cells.

258

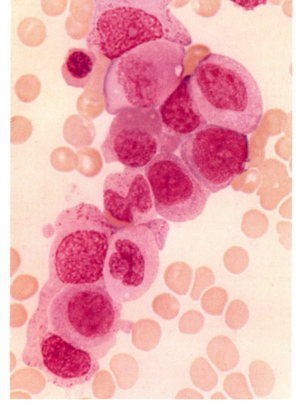

257

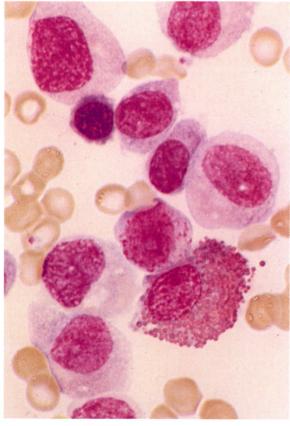

259

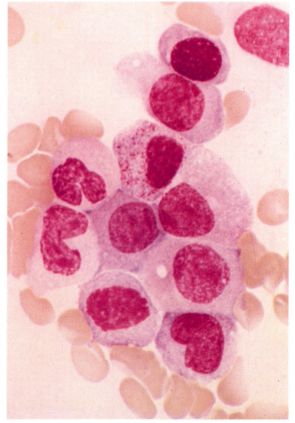

260. SB stain on the same marrow sample, showing strong sudanophilia, despite the lack of clear specific granules in the Romanowsky preparations.

261. High power view of the SB stain in this condition. The sudanophilic granules are of the same size as those seen in myelocytes with normal specific granules.

262. Dual esterase stain showing normal CE-positive granules in three of these unusual myelocytes, despite their poor granularity with the Romanowsky stains.

261

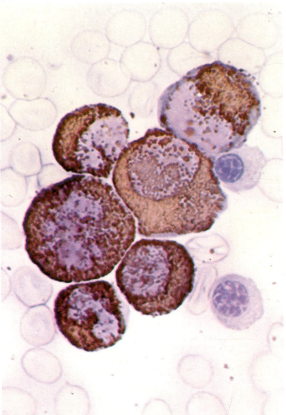

260

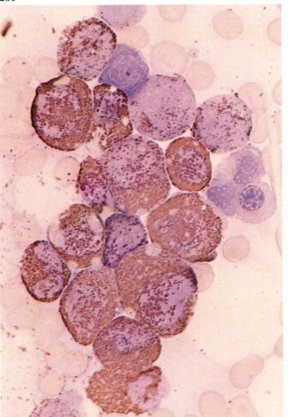

262

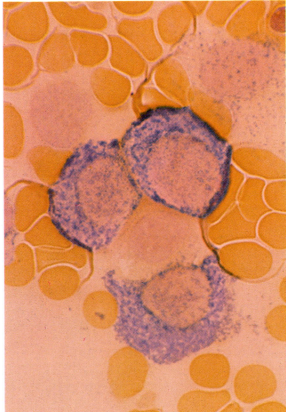

263–268. *Preparations from a case of juvenile CML with peripheral leucocyte count of 30 × 10⁹/l.*

263. Leishman stain of blood smear showing prominent monocytic component, several monocytes being vacuolated.

264. Leishman stain of bone marrow smear from the same patient, showing resemblance to CML, with various granulocyte precursors predominant.

265. SB stain on blood smear: coarse positivity in neutrophils and discrete scattered granules in several monocytoid cells.

264

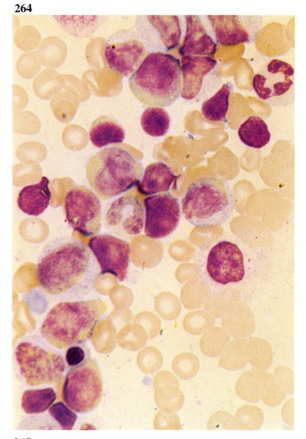

263

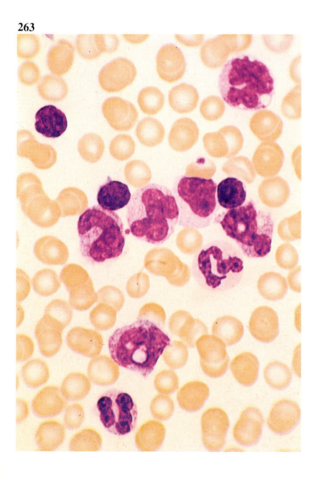

265

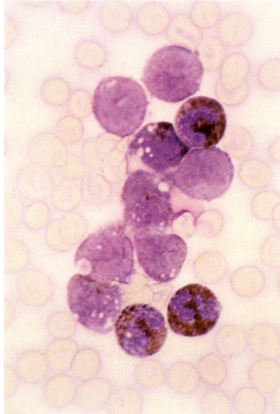

266. PAS stain on blood smear: a blast cell with monocytic PAS positivity pattern, two vacuolated monocytes and two neutrophil polymorphs with normal PAS reactions.

267. Acid phosphatase: three polymorphs and three monocytes in peripheral blood show moderately strong granular positivity.

268. Dual esterase: two polymorphs show CE positivity and six monocytes and precursors show moderate BE positivity.

267

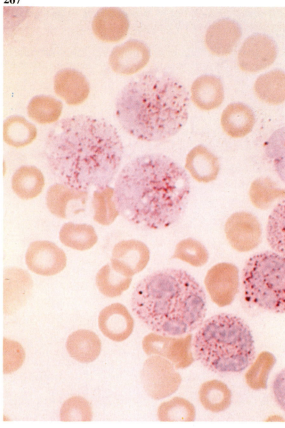

266

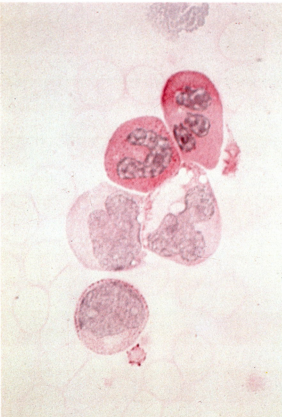

268

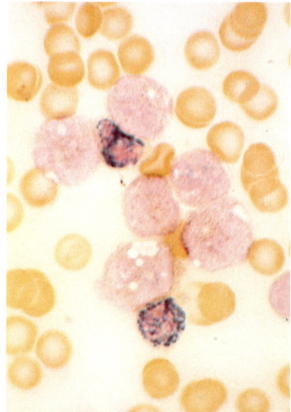

269. An 'LE' cell. The characteristic inclusion body of ingested nuclear material produced in leucocytes – especially polymorphs – as a result of the action of a factor present in the serum of patients with disseminated lupus erythematosus.

270 and 271. Further variations in staining and morphology of LE cells.

270

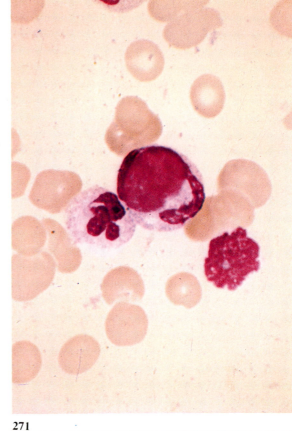

269

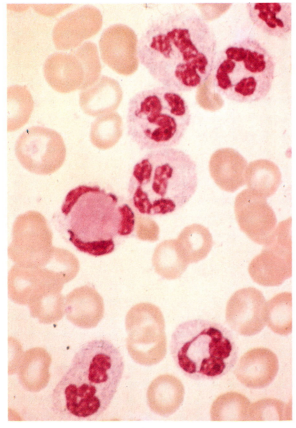

271

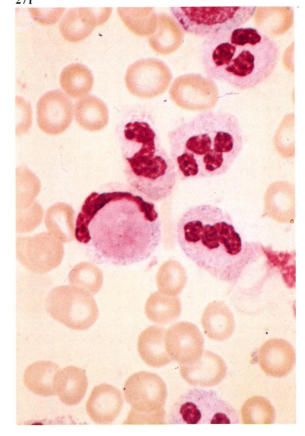

272. Another example of an LE cell with a large but partially divided inclusion.

273. A small LE cell inclusion almost surrounded by the nuclear lobes of a neutrophil polymorph.

274. A 'tart' cell, as commonly found in preparations made in the search for LE cells. The inclusion is usually, as here, in a monocyte, and consists of a cell nucleus, most often of a lymphocyte. The inclusion stains more deeply than does the LE body. These 'tart' cells are of no known pathological significance and must be distinguished from LE cells.

273
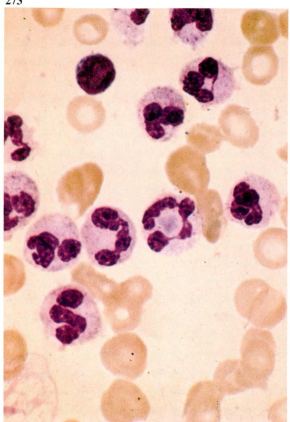

272
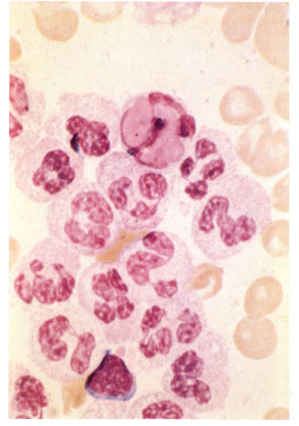

274
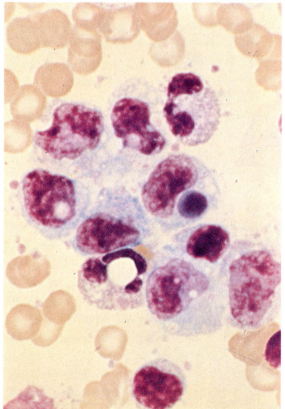

275. Pelger-Huet phenomenon. In the homozygous form the polymorphs show a single, rounded, dense nucleus. All the granulocytes are affected in the hereditary form of this disorder, but the change may affect some cells in myeloid leukaemia, giving a 'pseudo-Pelger' appearance.

276. Pelger-Huet phenomenon. Heterozygous form with most polymorphs showing a 'band' or bilobed nuclear structure. This appearance may occur as a familial abnormality, but may be mimicked in the 'pseudo-Pelger' polymorphs sometimes seen in acute and chronic myeloid leukaemias and in myelofibrosis.

   The example illustrated here is from a patient with chronic myeloid leukaemia. One of the cells shows twinning of bilobed pseudo-Pelger nuclei.

277. Coarse reddish-violet granules in leucocytes in Alder's anomaly, a familial disorder of leucocyte granularity. The appearance in the neutrophils resembles the 'toxic granules' commonly seen in leucocytosis of infection (as in **278** and **279**), but the accompanying granules in lymphocytes, sometimes within vacuoles, and occasionally showing a 'comma' shape, are strongly suggestive of Alder's anomaly.

276

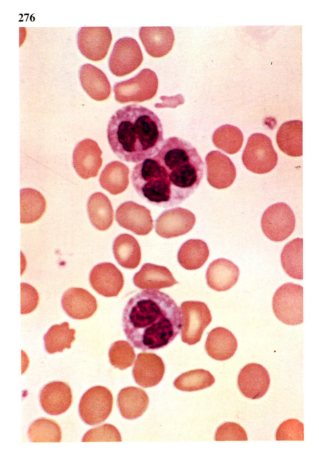

275

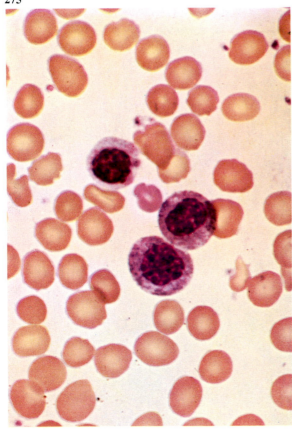

277

278 and 279. Examples of the blue-staining areas in the cytoplasm of neutrophil polymorphs sometimes seen in infections, especially pneumonia. They are known as Döhle bodies. Both polymorphs also show some coarse toxic granularity.

280. Further examples of Döhle bodies in neutrophil polymorphs; in this instance the specific granules are weak rather than coarse, and the Döhle inclusions more readily visible.

281 and 282. Examples of the May-Hegglin anomaly, an inherited disorder with basophilic inclusions, 2–5 μm in diameter, in granulocytes. The inclusions are larger than Döhle bodies and not related to infection.

279
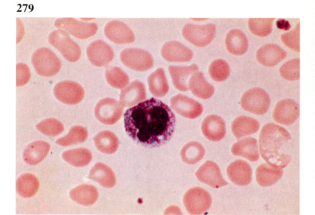

280
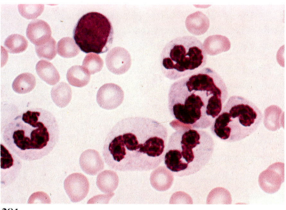

278
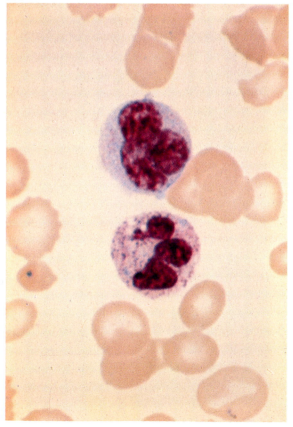

281
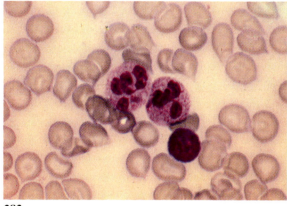

282
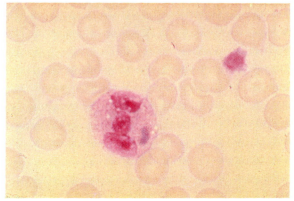

283 and 284. Further unusual inclusions, some 3–5 μm in diameter, weakly basophilic and apparently membrane-bound, in granulocyte precursors and in lymphocytes from a child with a transient pancytopenia and haemolytic syndrome. The material seems likely to be ribosomal protein.

285. Giant granulation, with both red and pale blue granules in a neutrophil polymorph. Similar giant granules in leucocytes of all kinds may be seen in the rare familial Chediak-Higashi-Steinbrinck anomaly.

286. Conspicuous inclusion particles, probably carbon, in a peripheral blood monocyte.

284

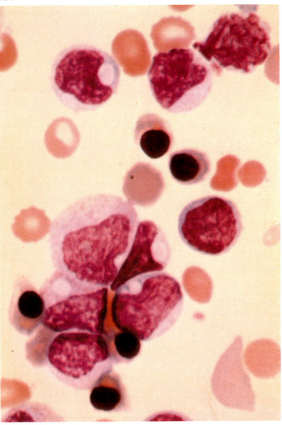

283

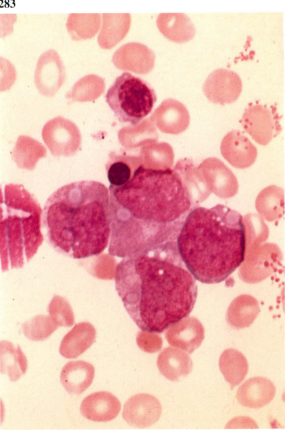

285

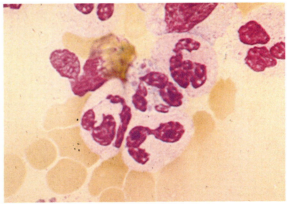

286

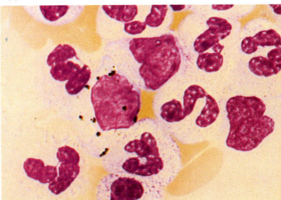

287. A monocyte with some reddish granules in the grey cytoplasm, together with a lymphocyte, a stab cell, and two neutrophil segmented polymorphs.

288. Two monocytes, one with vacuoles, and a lymphocyte.

289. Two monocytes and a stab cell. The monocytes show moderate cytoplasmic granularity and typical nuclear shapes.

288

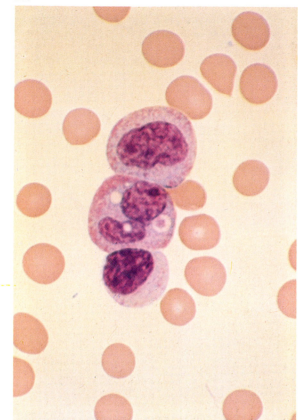

287

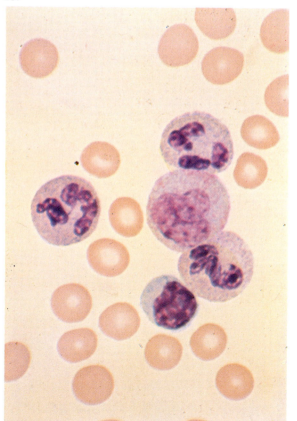

289

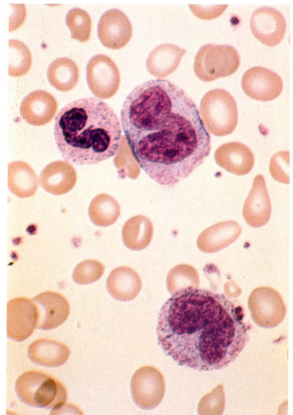

290. Peripheral blood monocyte and a neutrophil polymorph, each with little cytoplasmic granularity.

291. Promonocytes from a case of acute 'myeloid' leukaemia with predominance of the monocytic series. The cells, though nucleolated, show nuclear indentations and a cytoplasmic colour tending towards the monocytic grey rather than the basophilia of less differentiated cells.

Pure 'monoblastic' or 'monocytic' leukaemias are very uncommon; there is nearly always some granulocytic element present, so that the name 'myelomonocytic' is often used, but as in this case, cells of the monocyte series may greatly predominate, and it is convenient to retain the term acute monocytic leukaemia for such cases.

292. Sudan black B staining in the same case as in **291.** The promonocytes show discrete scattered granules of positivity, here chiefly confined to the cytoplasm, but not densely clumped as in granulocyte precursors.

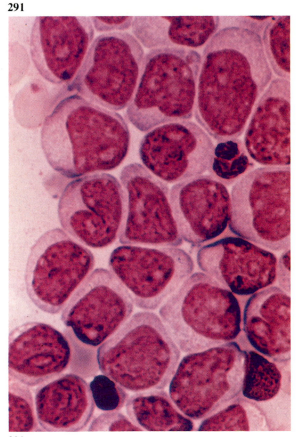

290

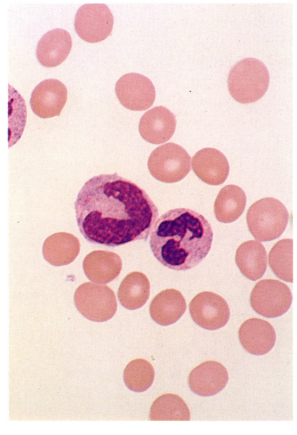

292

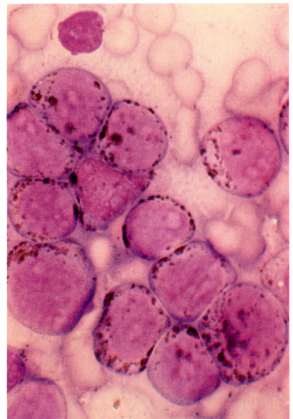

293. Acute monocytic leukaemia, with monocyte precursors of variable morphology and state of maturity. This field illustrates also a scattering of plasma cells. An increase in plasma cells, generally focal or patchy in distribution, is commonly observed in acute leukaemias of any kind and is also seen when other malignant processes invade the marrow.

294. Sudan black reaction in the same case. A single strongly sudanophilic polymorph contrasts sharply with the monocyte precursors, with their reactions ranging from negative to moderately strong positivity of the discrete scattered granule type.

295. Three primitive cells from an acute leukaemia with minimal morphological signs of differentiation in Romanowsky preparation. The Sudan black reaction, shown here, has the typical distribution of positive granules characteristic of the monocytic series. The primitive blast cells are therefore monoblasts.

294
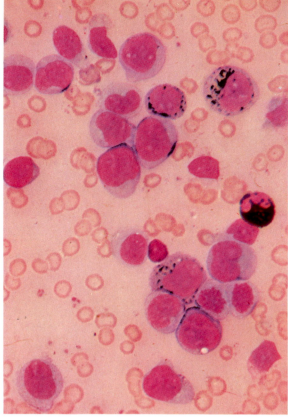

293
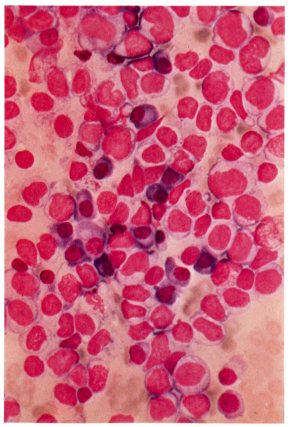

295
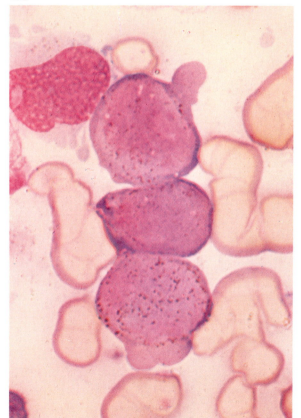

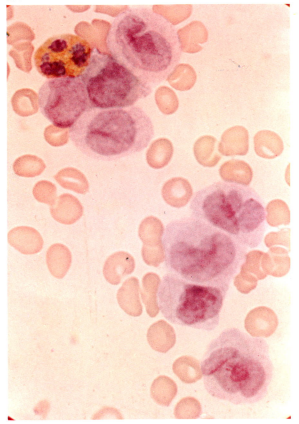

296. Acute monocytic leukaemia: peroxidase reaction. Monocytes and their precursors are generally negative for peroxidase, but may show a faint localised cytoplasmic reaction. Auer rods, as shown in the upper left corner, are strongly peroxidase positive. Their presence confirms the myelo-monocytic nature of the disease, since Auer rods probably do not occur in the monocyte line but are confined to granulocyte precursors, being derived chiefly from primary, azurophil, granules.

297. Another example of peroxidase staining in an acute monocytic leukaemia. The monocytes appear rather more differentiated than in the previous example, and have a negative reaction.

298. Sudan black reaction in the same case: the monocytes mostly show discrete scattered granules. A myeloblast with two positive Auer rods is also seen.

296

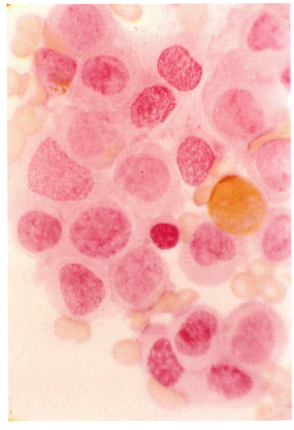

298

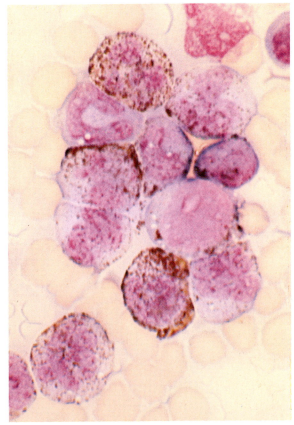

299. A group of monocyte precursors from an acute monocytic leukaemia.

300. The same field, consecutively stained with the PAS reaction. Monocyte precursors show considerable variability in their positivity to PAS, from completely negative reactions to coarsely granular, heavy positive ones. Here the cells show a mixture of diffuse cytoplasmic staining and fine granules.

301. A series of monocytes and precursors from an acute monocytic leukaemia, to illustrate the range of variation encountered in PAS reactivity in this cell series.

302. Very strong PAS positivity with coarse blocks in monocytes from a further case of myelomonocytic leukaemia with predominance of well differentiated monocytes.

301

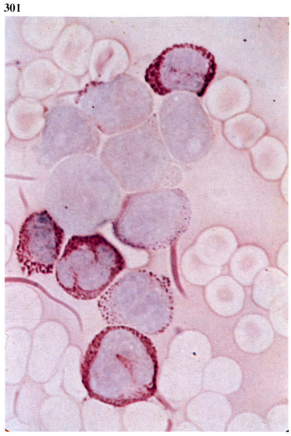

302

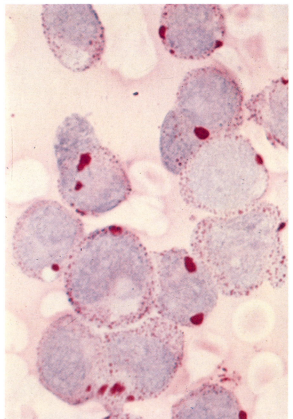

299

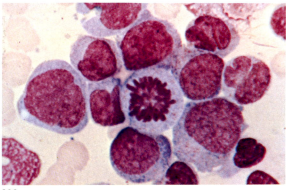

300

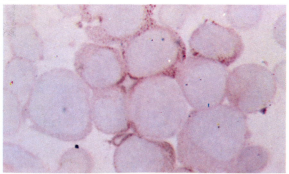

303. Acid phosphatase reaction in acute monocytic leukaemia: most cells show strong and coarsely granular positivity.

304. Dual esterase reaction in acute monocytic leuk-aemia: monocyte precursors show strong butyrate esterase (BE) positivity in this case, while a granulocyte shows strong chloroacetate esterase (CE) positivity.

305. Dual esterase reaction in a case of acute myelo-monocytic leukaemia. In this instance the monocyte precursors show only weak reactions with scanty scattered granular BE positivity. Two myeloblasts show chiefly CE positivity but some granules of BE reaction are also detectable.

304

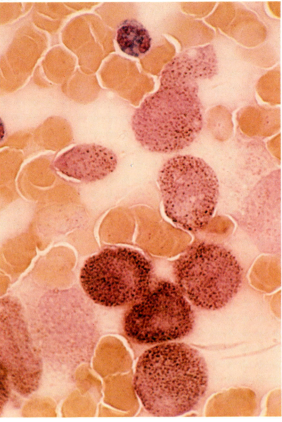

303

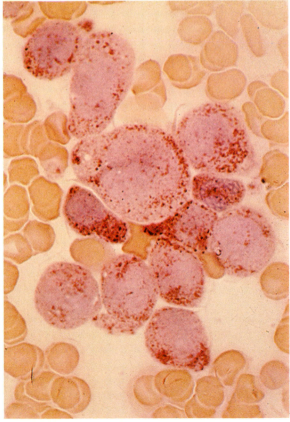

305

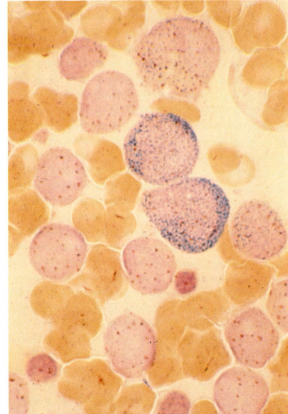

306–308. Romanowsky, PAS and dual esterase reactions on the cells of an acute myelomonocytic leukaemia with predominantly monocytic morphology and typically monocytic mixed diffuse and granular PAS reaction, but with CE rather than BE positivity in the monocytes. This is a very uncommon picture but illustrates the potential variability of monocytic esterase content.

307

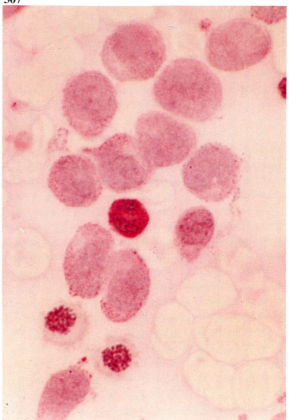

306

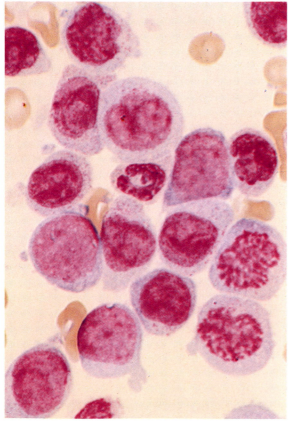

308

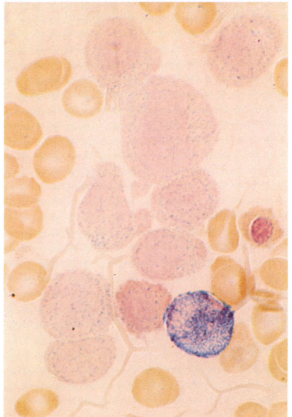

309. An acute myeloid leukaemia with mixed granulocyte and monocyte precursors; when admixture is marked, as in this case (suggested by the Romanowsky stain and confirmed by the Sudan black), the term myelomonocytic is conveniently applied.

310. Sudan black reaction from the same case; the localised cytoplasmic positivity in the early granulocyte precursors contrasts with the discrete scattered granule pattern of the monocyte precursors.

311. Dual esterase reaction in a similar case, with typical BE reaction in the monocyte precursors and CE reaction in granulocyte precursors.

310

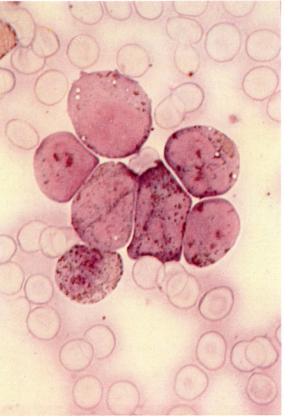

309

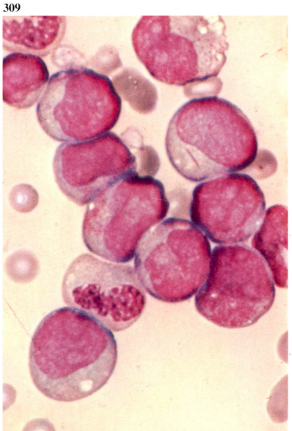

311

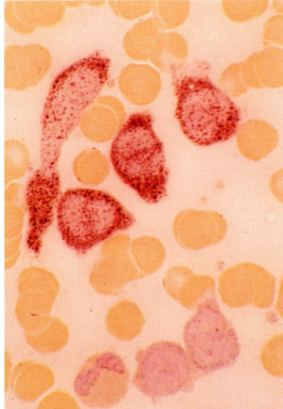

312. An acute 'myeloid' leukaemia with mixed erythroid, granulocytic and monocytic precursors – an 'erythro-myelo-monocytic' leukaemia. Such mixed leukaemias are quite common, an observation which supports the concept of a 'myeloid' stem cell with multiple potentialities.

313. The Sudan black reaction helps to differentiate the negative pro-erythroblasts from the myeloblasts with dense localised positivity and the monoblasts with scattered granules.

314. A general view of a bone marrow smear from the same case as shown in the figures above, stained by the PAS reaction. The diffuse and finely granular positivity of granulocyte precursors, increasing in intensity with increasing maturity, is shown; the strong positive reaction in some erythroblasts is very characteristic of erythraemic involvement in a mixed leukaemic process.

313
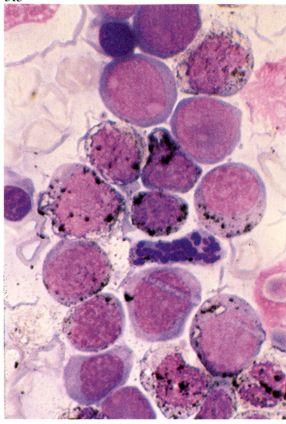

312
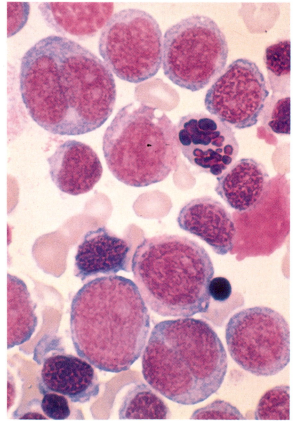

314
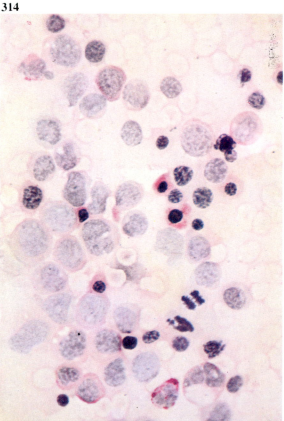

315. Acute 'myeloid' leukaemia, with mixed proliferation of erythroid and granulocyte precursors – an 'erythromyeloid' leukaemia. Most cases of erythraemic myelosis have from the earliest stages some component of granulocytic or monocytic cell line involvement, which, though it may be minimal at first, comes often to dominate the picture eventually.

316. A higher power view of primitive cells from the same case; they are not altogether easy to classify on the Romanowsky preparations and cytochemical assistance is required.

317. The Sudan black reaction on a smear from the same case of erythroleukaemia as shown in the last two figures. Erythroid precursors are negative, while granulocyte precursors show typical coarse localised positivity.

316

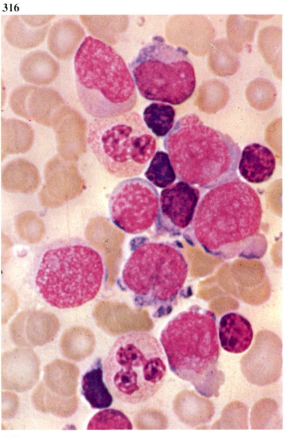

315

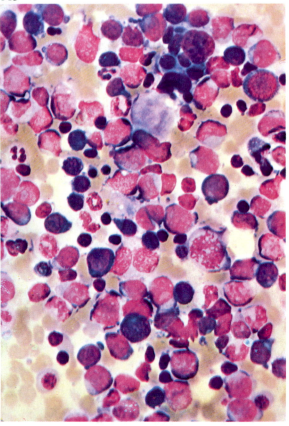

317

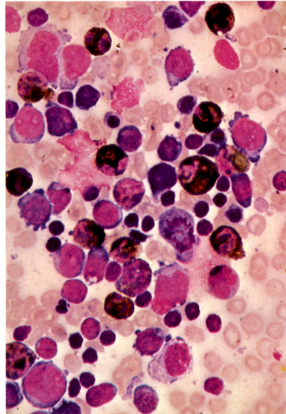

318. PAS reaction in the same case. Strong positivity in red cell precursors at different stages of maturity is conspicuous, while myeloblasts and promyelocytes show their customary negative reactions or faint diffuse tinge of the cytoplasm.

319. Acid phosphatase reaction in mixed erythromyeloid leukaemia. Coarse positivity, especially paranuclear, in all cells present.

320. Dual esterase reaction in the same case. An early granulocyte precursor shows CE positivity and the erythroblasts fine granular positivity of mixed BE and CE type.

319

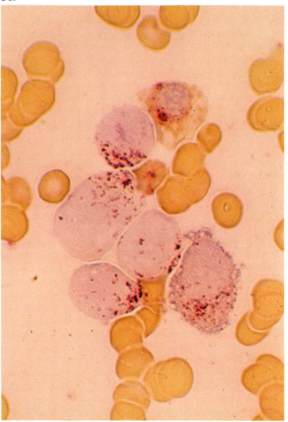

318

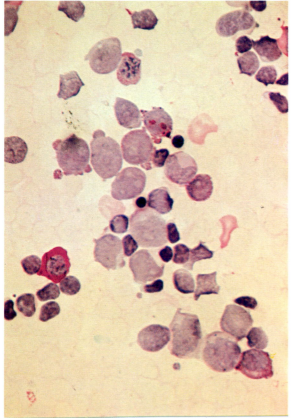

320

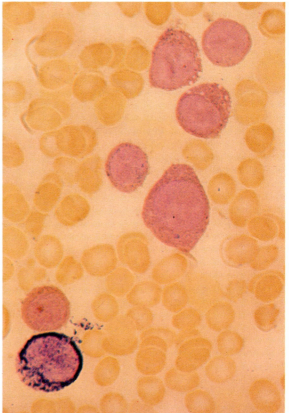

321. A normally granular mature megakaryocyte, with minimal platelet formation at the periphery.

322. A megakaryocyte with disrupting cytoplasm which has been actively forming platelets.

323. A poorly granular megakaryocyte with minimal platelet formation.

324. A pair of megakaryocytes, very actively releasing platelets, and almost devoid of cytoplasm.

325. Stages in the formation of megakaryocytes. The large cell with three nuclei and fragmenting cytoplasm might be called a 'promegakaryocyte', and the accompanying primitive cell, like an unusually large myeloblast, may be a 'megakaryoblast'.

323

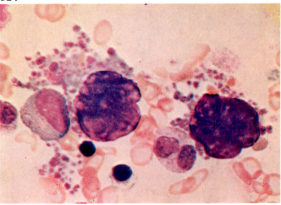

324

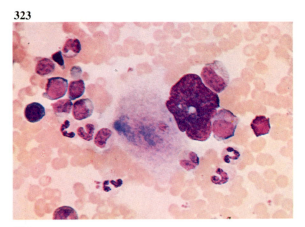

321

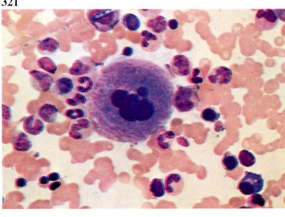

322

325

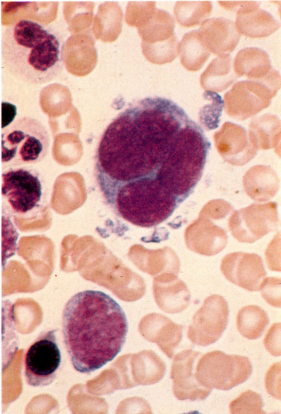

326. Mature megakaryocytes with strong granularity of their cytoplasm.

327. A megakaryocyte fragment in the peripheral blood (from blastic crisis in chronic myeloid leukaemia).

328. A giant platelet of snake-like form, beside various red cell precursors in the bone marrow of a patient with iron deficiency anaemia.

327

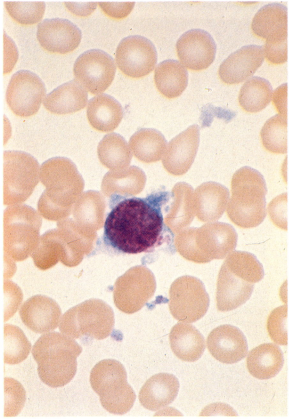

326

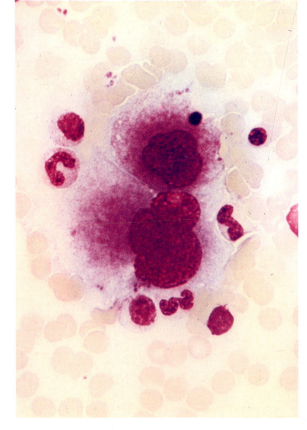

328

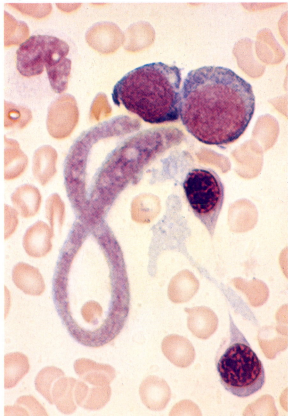

329. Phagocytosis of platelets by neutrophil polymorphs in an autoimmune disorder with circulating immunoblasts, two of which are seen in this field.

330. PAS reaction in a normal megakaryocyte. Diffuse cytoplasmic positivity, weak in intensity, is accompanied by occasional strongly positive glycogen inclusion bodies.

331. When platelet formation is active or imminent, a peripheral rim of denser positivity may also be observed. In this low power view of normal marrow, the megakaryocyte is surrounded by other marrow cells which show the increasing positivity with greater maturity in the granulocyte series. Erythroid precursors are negative.

330

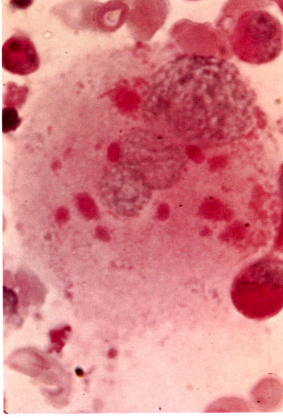

329

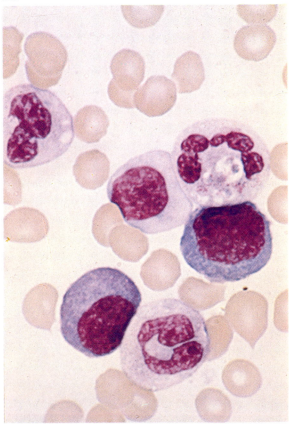

331

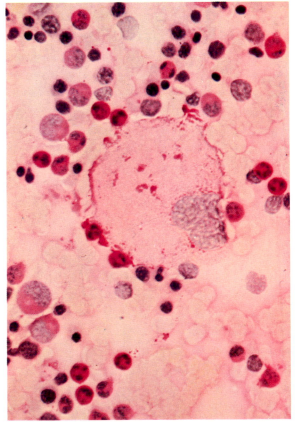

332. Numerous megakaryocytes without peripheral platelets from the marrow of a patient with chronic idiopathic thrombocytopenic purpura (ITP).

333. Megakaryocytes from another patient with chronic ITP to show the intense accumulation of glycogen inclusion bodies sometimes seen in this disorder.

334. Megakaryocyte from the same patient as in **333** but after splenectomy (with good response), to show the disappearance of glycogen inclusion bodies in the PAS stain.

333

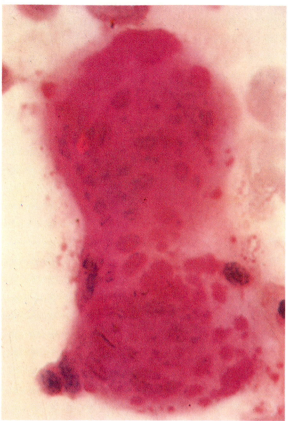

332

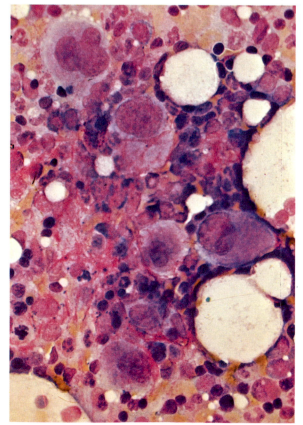

334

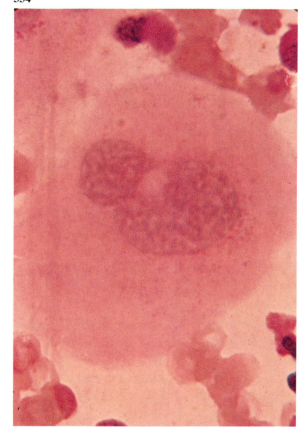

335. Acid phosphatase reaction, showing coarse granular scattered positivity in a megakaryocyte; two myelocytes show a few granules as does a late normoblast, and a lymphocyte is negative.

336. Dual esterase reaction, showing weak granular BE positivity in a megakaryocyte: three granulocytes show CE positivity, a monocyte is BE positive, and two lymphocytes are negative.

337. α-Naphthyl acetate esterase (AE) reaction in a megakaryocyte – the positivity is much stronger and denser than with butyrate as substrate, a characteristic feature of megakaryocytes.

336

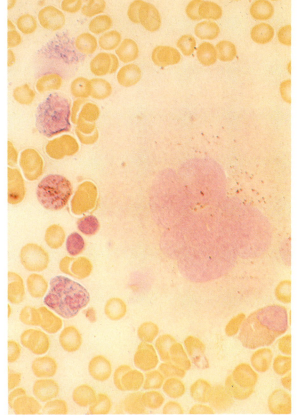

335

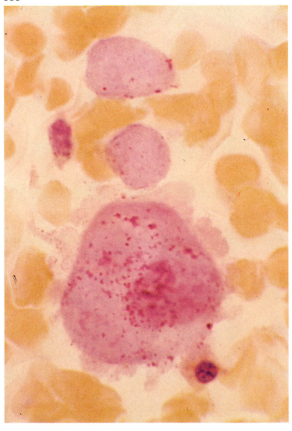

337

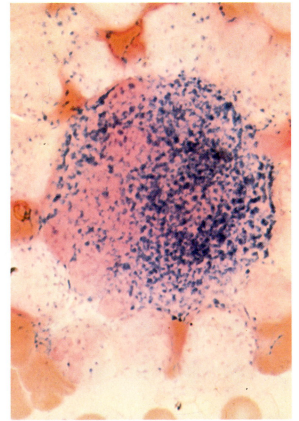

338 and 339. Gross phagocytosis of an erythroblast clump by one megakaryocyte and phagocytosis of other cellular debris and red cells by other megakaryocytes in a myeloproliferative disorder with megakaryocytic hyperplasia.

340. PAS stain in the same case, showing coarse PAS positivity in two megakaryocytes, each of which contains ingested leucocytes.

339

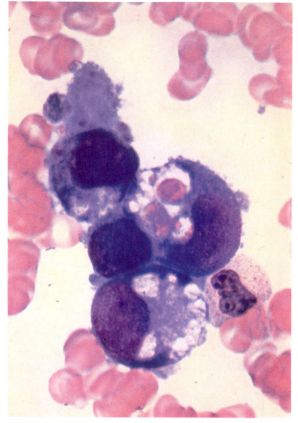

338

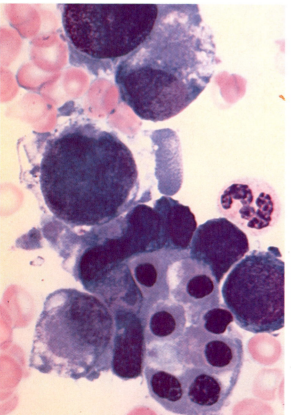

340

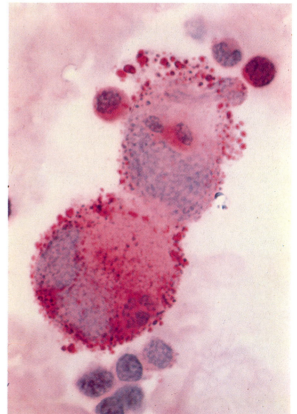

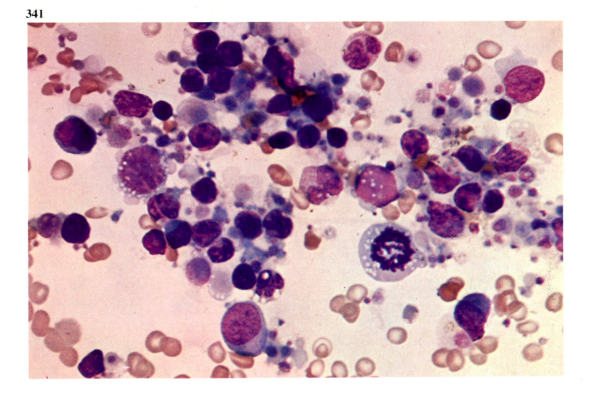

341. Numerous megakaryocyte fragments and probable megakaryocyte precursors in the peripheral blood of a patient with an acute 'myeloid' leukaemia with mega-karyocytic preponderance – the term acute megakaryo-cytic myelosis may be applied. Large and fully developed megakaryocytes are not present, but multiple megakary-ocyte fragments or abnormal small cells of that series, where the usual polyploidy has not occurred.

342. A further field from acute megakaryocytic myelosis. The nucleolated primitive cells have a similarity to myeloblasts, but may be 'leukaemic' megakaryoblasts.

342

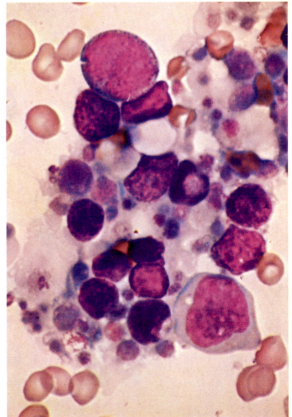

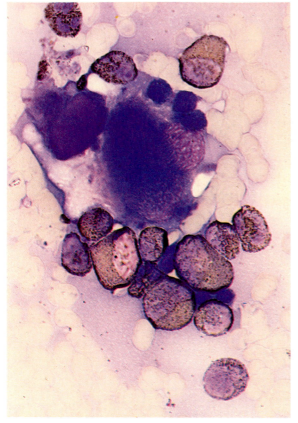

343. Sudan black reaction in the same case. A megakaryócyte with three phagocytosed lymphocytes or erythroblasts is negative, as is the neighbouring megakaryoblast.

344. PAS reaction in acute megakaryocytic myelosis. The strong and coarse irregular positivity in the 'leukaemic' megakaryocytic cells resembles that in platelets. The diffuse tinge in certain precursors does not allow a clear distinction from myeloblasts to be made.

344

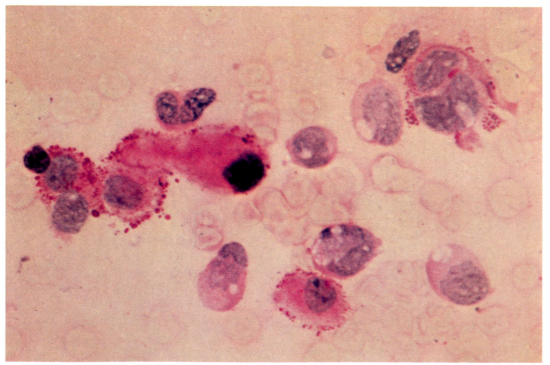

345–356. *Early cytological signs of emerging myeloid haemopoietic activity during remission development in acute leukaemia. The features illustrated are usually found in marrows which otherwise remain hypoplastic following cytotoxic therapy.*

345. A plasma cell, a reticulo-endothelial cell (RE cell, macrophage) and the granular, platelet-forming cytoplasm of a megakaryocyte.

346 and 347. Young megakaryocytes, showing new platelet formation at sites where the primitive cytoplasmic basophilia is first lost.

346

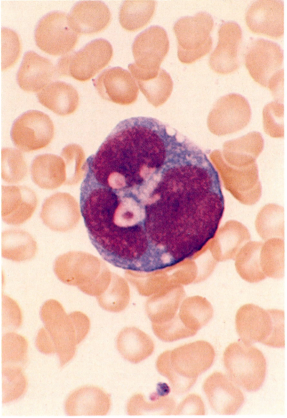

345

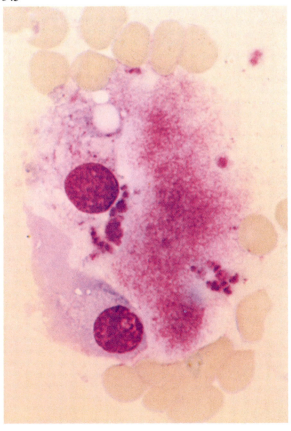

347

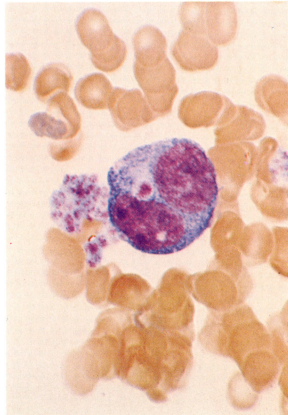

348. A more mature megakaryocyte, but still with patches of residual basophilia.

349. RE cells may appear relatively numerous at this phase – the marrow fleck here is composed almost entirely of RE cells. Elsewhere the marrow showed gross hypoplasia, yet remission ensued within a few days.

350. A higher power view of the cells in this fleck – the nuclear pattern is characteristic of RE cells.

349

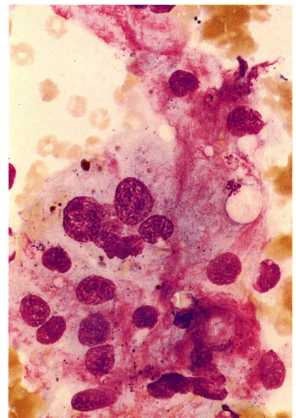

348

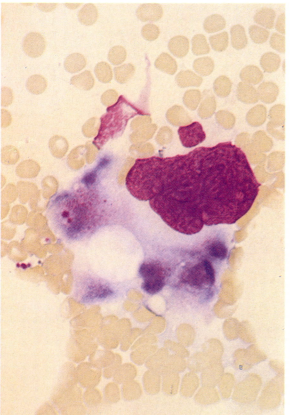

350

351. A group of typical remission-associated cells: plasma cells, RE cells, megakaryocyte-platelets, a single blast cell, granulocyte precursors and a giant megaloblast following antimetabolite chemotherapy.

352. A myeloblast, two coarsely granular promyelocytes (as typically found in early remission) and a metamyelocyte.

353. A similar sequence of granulocyte precursors together with stab and segmented neutrophils of normal appearance.

352

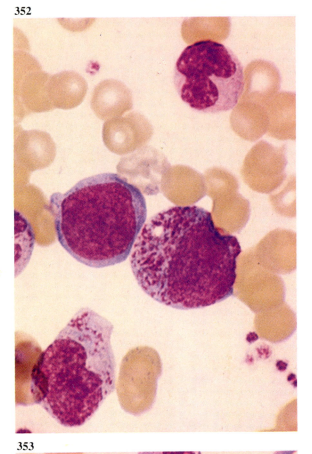

351

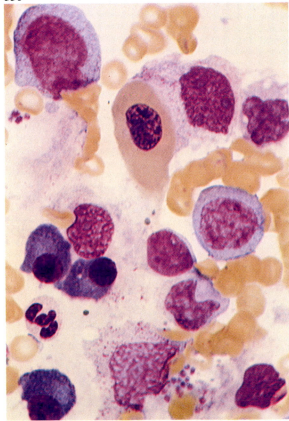

353

354–356. Examples of giant or multilobed neutrophil metamyelocytes or polymorphs in bone marrow aspirates taken during early stages of remission emergence in acute leukaemia. Figure **356** also shows an activated lymphocyte or immunocyte approaching plasma cell morphology. Pathological though the multilobed neutrophils appear, they do seem frequently to precede remission development.

355

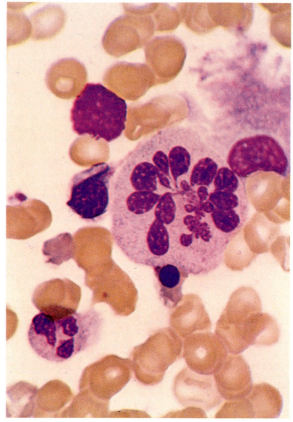

354

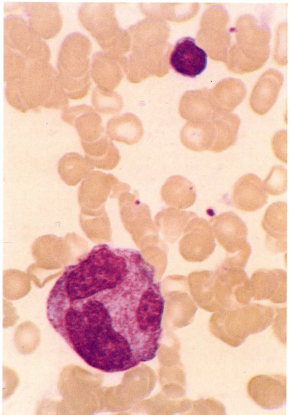

356

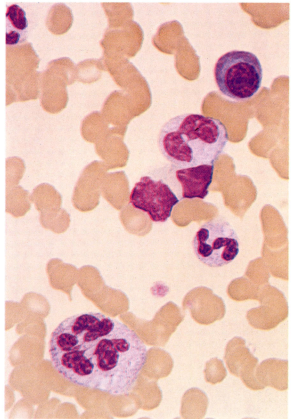

Part 3
Lymphocytes, plasma cells and their derivatives and precursors in blood and bone marrow
Normal and abnormal forms

Lymphocytes are formed chiefly in the lymph glands, spleen, Peyers patches and other nodal sites – among which must be included nodules of lymphocytic nature in the bone marrow. Precursors, lymphoblasts and prolymphocytes, are difficult to recognise in normal marrow aspirates, however, and are illustrated from pathological conditions, principally acute lymphoblastic and prolymphocytic leukaemia, where they become common in the marrow and sometimes also in the peripheral blood.

While the earliest stages of lymphoid cell development from the probable common stem cell shared with the myeloid series are still uncertain, it seems likely that a functional stem cell for the lymphocytic cell series exists which gives rise to two main sorts of lymphocyte, T cells and B cells. The former have a special role in cell-mediated immune responses and may have helper (Tμ subset) or suppressor (Tγ subset) functions in relation to B-cell activity. B cells are largely responsible for humoral immune responses. These cells, and smaller groups of lymphocytes which seem to fall into different, less clear, categories are distinguishable by immunological means but also have certain morphological and cytochemical differences which are illustrated in this section, and which are sometimes also manifest in neoplastic states.

Under stimulation by phytohaemagglutinin normal T lymphocytes may undergo, in vitro, a process of dedifferentiation or transformation to produce primitive cells, with basophilic cytoplasm and leptochromatic nucleolated nuclei, able to divide. A similar transformation of B cells may occur following antigenic stimulation, and the primitive cells resulting may undergo further cytological transformation to give rise to plasma cells.

The close relationship of lymphocytes and plasma cells in their functional contribution to the immune response is not paralleled by any remarkable mixed proliferations of the two morphologically distinguished cell lines in leukaemic states. Lymphoblastic and lymphocytic leukaemias may have a component of plasma cells present, but it is rarely conspicuous and often absent. Lymphoblastic leukaemias may be of early non-T non-B origin (either common (c)ALL or nullALL according to immunological criteria), or may show evidence of T-cell features, or, rarely, of B-cell features. Chronic lymphocytic and prolymphocytic leukaemias are mostly of B cells and only rarely of T cells. Hairy cell leukaemia is probably also a B-cell neoplasm, but Sezary's syndrome seems mostly to involve T cells.

Myelomatous proliferation, whether as solitary nodules, multiple bony lesions, or disseminated marrow involvement, with spread to peripheral blood, has no consistent or frequent element of lymphocytic hyperplasia but appears to be a tumour of the terminal plasma cell phase of the B-cell line. In macroglobulinaemia the neoplastic cells are activated IgM-secreting lymphocytes or lymphoplasmacytoid cells.

Morphological variations in both lymphocytes and plasma cells cover a much wider range than among granulocytes. In infections, especially those due to viruses, lymphocytes show activated or immunoblastic features intermediate between mature and primitive cells, with increase in cytoplasmic basophilia, and sometimes the appearance of visible nucleoli. Even in normal blood there may be considerable variations in the size of lymphocytes and the amount of cytoplasm. Cytoplasmic vacuoles and inclusions are not infrequent. Fine or coarse granules and even blocks of glycogen may occur in lymphocytes and lymphoblasts, and often appear especially conspicuous in leukaemias or other lympho-proliferative states.

Plasma cells show even more abundant cytological variants, mostly connected, there is little doubt, with their production of immunoglobulins and the accumulation of part or the whole of these protein molecules in various morphological guises in the cytoplasm. Here again the cytological variants are most conspicuous in multiple myeloma, the plasma cell equivalent of leukaemia, but any of them may be seen occasionally in plasma cells from other conditions and even in normal marrows, and none appears to be specifically confined to myeloma.

Lymphoma, a malignant disease arising in lymph nodes and involving cells of more or less mature degree in the lymphocytic line, commonly extends at some stage to involve the bone marrow and to liberate abnormal lymphocytes in the peripheral blood. If the cell type is mature the disease is cytologically indistinguishable from chronic lymphocytic leukaemia, and if very primitive, as it may be especially in children, it is equally indistinguishable from acute T or B cell lymphoblastic leukaemia. In some cases, however, abnormal 'lymphoma cells', unlike either mature lymphocytes or the lymphoblasts of acute leukaemia but with characteristics intermediate between the two, may be seen. These cells are nearly always of the B-cell line and include neoplastic variants of the centrocytes, centroblasts and immunoblasts present in lymph nodes and illustrated more fully in Part 5.

The cytology and cytochemistry of all these normal and abnormal variants as seen in blood and marrow are illustrated in the following pages.

357. A lymphoblast with a mature lymphocyte and a stab cell, from the peripheral blood of a patient with acute lymphoblastic leukaemia (ALL).

358. Lymphoblasts from the common type of non-T non-B acute lymphoblastic leukaemia (cALL), showing the high nuclear-cytoplasmic ratio and occasional nuclear cleavage – Rieder cell formation. There are a few mature lymphocytes and some intermediate 'prolymphocytes' present.

359. Another example of cALL. There are two granulocytic cells present – a metamyelocyte and an eosinophil myelocyte – and also some cells with denser nuclei maturing along the lymphocytic line. This field illustrates the difficulty of certain allocation of acute leukaemia to a specific type on the appearances in Romanowsky stains alone. Cytochemical staining in this case gave the typical findings shown in the following figures and established the diagnosis unequivocally.

358
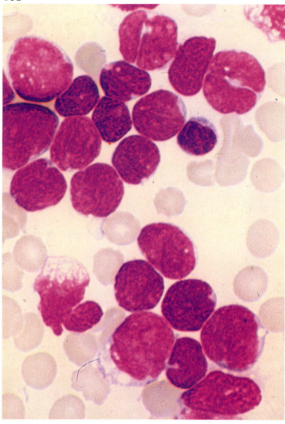

357
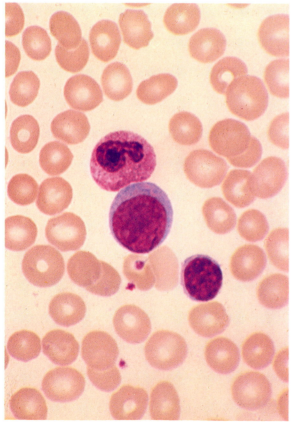

359
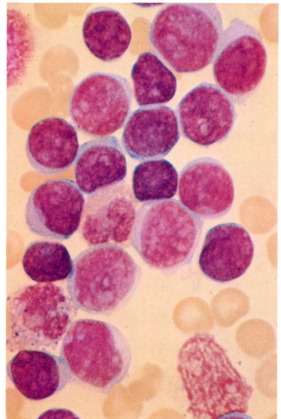

360. Sudan black stain in cALL. There is a myelocyte with normal positivity present (for contrast) but the lymphoblasts and occasionally more mature cells of the lymphocyte line are uniformly negative. As well as the high nuclear-cytoplasmic ratio occasional vacuolation can be seen.

361. Peroxidase reaction in cALL. Lymphoblasts are uniformly negative. A polymorph present is normally positive.

362. PAS stain in acute lymphoblastic leukaemia (cALL). The lymphoblasts mostly show numerous fine and coarse granules of positivity. Note that the erythroblasts present are negative – unlike those forming an erythraemic component in mixed myeloid leukaemias. A single myelocyte shows normal diffuse but weak PAS positivity.

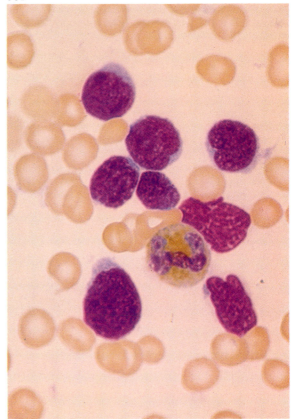

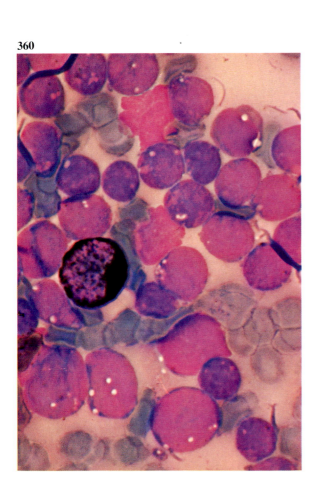

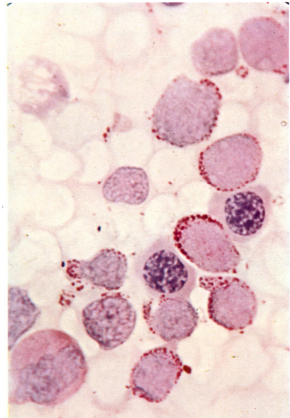

363. Another example of PAS staining in acute lymphoblastic leukaemia. In this case some cells contain blocks in addition to coarse granules.

364. A group of lymphoblasts from cALL. The cytoplasm shows frequent irregularities of staining and occasional vacuoles, especially in cytoplasmic buds.

365. The same field as above, consecutively stained with the PAS reaction. Most of the lymphoblasts contain coarse granules of PAS positive material, often coinciding with vacuoles or areas of lighter staining within the cytoplasm and especially in cytoplasmic protruberances as shown in **364**.

364

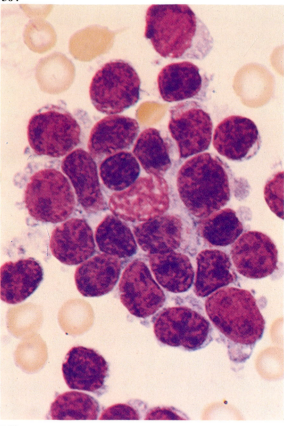

363

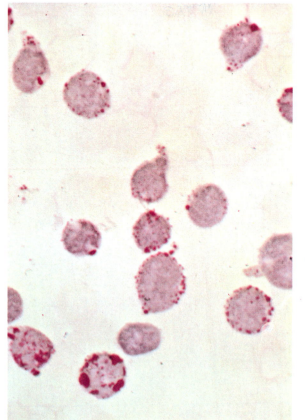

365

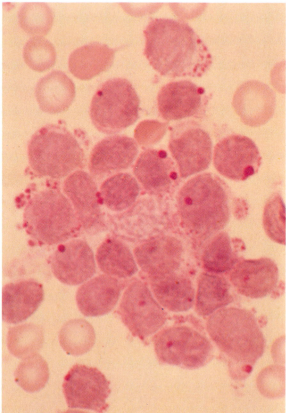

366. PAS reaction in a further case of cALL with minimal positivity. Even such almost entirely negative lymphoblasts commonly show some granular positivity in a very occasional cell. Weak reactions are perhaps commoner in T-cell cases, but this is not a very useful discriminant feature as much variation exists.

367. Acid phosphatase reaction in the same case, showing little positivity, typically for cALL. There is a normally positive polymorph present.

368. Dual esterase reaction with weak scattered BE positivity in lymphoblasts of cALL. This pattern is usual in non-B non-T cases.

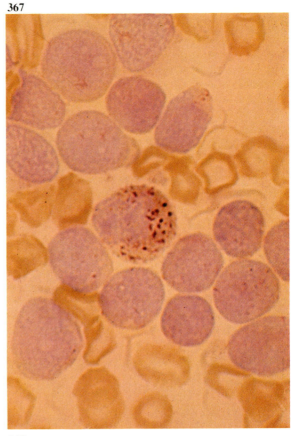

367

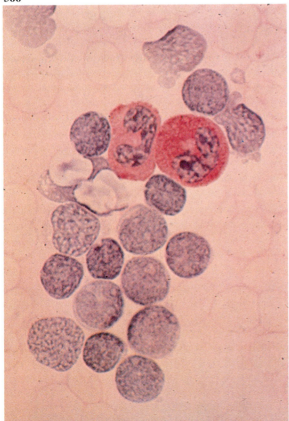

366

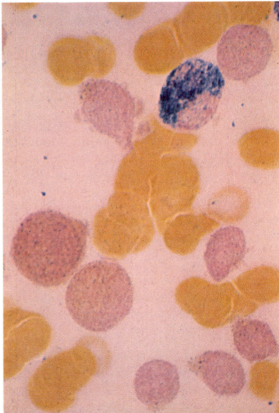

368

369. Acid phosphatase reaction in a T-cell ALL showing the strong localised paranuclear positivity in most lymphoblasts characteristic of this ALL variant. Such positivity is much less commonly found in cALL and null-cell ALL and is absent in only a small minority of cases of T-ALL.

370. Dual esterase reaction, showing localised BE positivity in lymphoblasts of T-cell ALL.

371. Scattered BE positivity in lymphoblasts, with a tendency to localisation to one side of the nucleus though less striking than in **370**. This pattern suggests a T-cell lineage, but the cells are in fact from a cALL case.

370

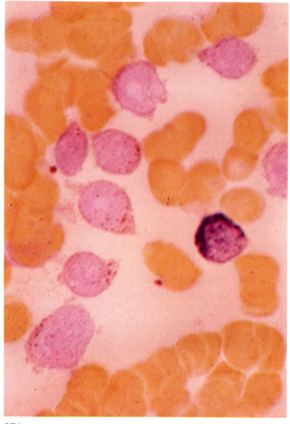

369

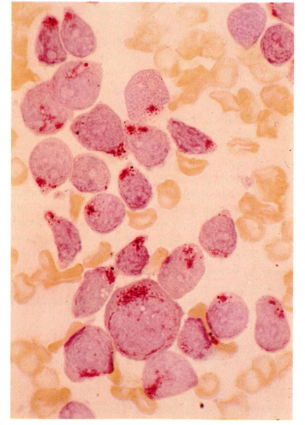

371

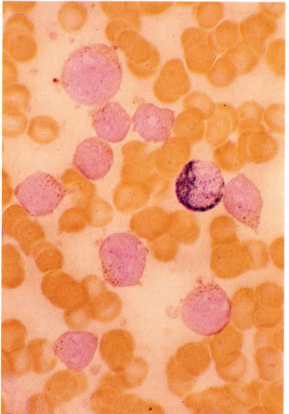

372. cALL plus normoblastic hyperplasia. The field shows some 20 blasts and 16 normoblasts and an RE cell. This type of mixed cytology is seen in phases of emerging remission or relapse in ALL.

373 and 374. *cALL plus marked eosinophilia. Curiously, eosinophilia is at least as common in ALL as in AML.*

373. Seven lymphoblasts with seven eosinophil or eosinophil precursors and a late normoblast from the marrow of a patient with cALL.

374. A lymphoblast, a lymphocyte and nine pathological eosinophils from the peripheral blood of the same case of cALL with gross eosinophilia. The eosinophil granules are defective in number and smaller than normal in size.

373

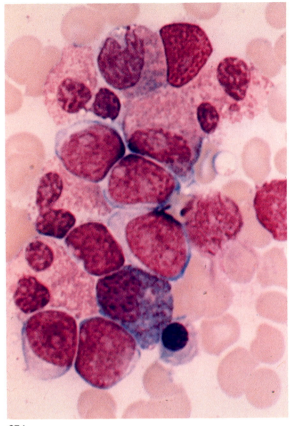

372

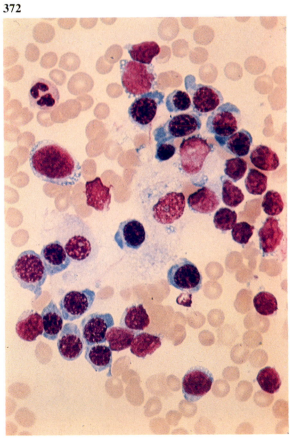

374

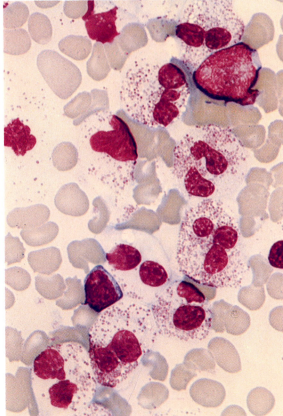

375. Three mature lymphocytes, together with a basophil, an eosinophil, and two segmented neutrophil polymorphs.

376. A binucleated lymphocyte. Such cells may rarely be seen in normal blood and are not, as at one time thought, suggestive of an irradiation effect.

377. A lymphocyte with more cytoplasm than those shown above, and with azurophil granules, possibly a suppressor T cell (Tγ). A neutrophil polymorph is also present.

376

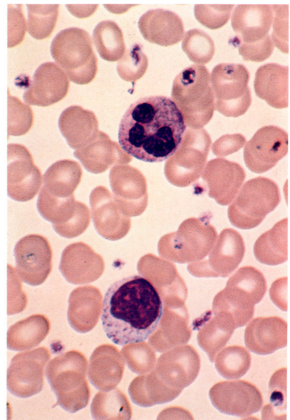

375

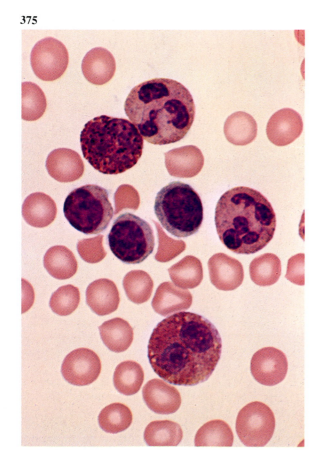

377

378–383. *Lymphocyte subsets and their cytochemical patterns with acid phosphatase and butyrate esterase reactions.*

378. Variations in lymphocyte cytology in a normal buffy coat smear. One of the five lymphocytes has little cytoplasm and is probably either a B or a Tμ cell, one has some marginal hair-like processes, not infrequently seen in normal lymphocytes and not to be confused with the appearance of hairy cells (see **441** to **443**), and the remaining three lymphocytes have relatively large amounts of cytoplasm with a few scattered granules, and are probably Tγ cells.

379. Acid phosphatase reaction in buffy coat cells. Two of the three lymphocytes show localised coarse dot positivity characteristic of the Tμ cell subset, the third is larger with more cytoplasm and with a little scattered positivity only and is probably a Tγ cell.

380. Acid phosphatase reaction in peripheral blood lymphocytes: the proportion of cells with the localised positivity shown here in three lymphocytes closely parallels the Tμ cell proportion by surface marker techniques. This field also shows granular acid phosphatase positivity in two neutrophils, a basophil and an eosinophil.

379

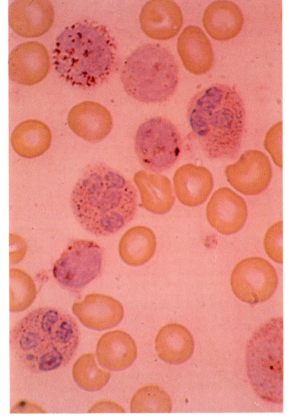

378

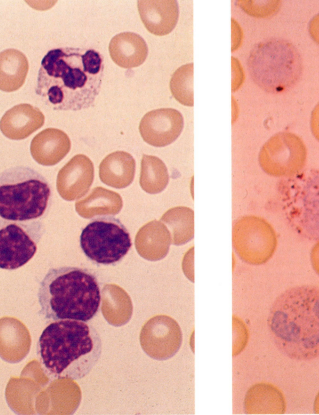

380

127

381. Dual esterase reaction. Localised T-cell positivity in one lymphocyte, probably a Tμ cell, and weak granular BE positivity in a second. This may be a null cell or perhaps a cell of the Tγ subset. There are four CE positive polymorphs and a BE reacting monocyte.

382. Two normally reacting (CE positive) polymorphs, a BE positive monocyte and four lymphocytes, two showing the localised dot-like Tμ-subset type positivity, one small, negative, probable B cell and a larger lymphocyte with ample negative cytoplasm, probably of the Tγ subset.

383. A monocyte and a polymorph, together with two lymphocytes of the Tμ subset.

382

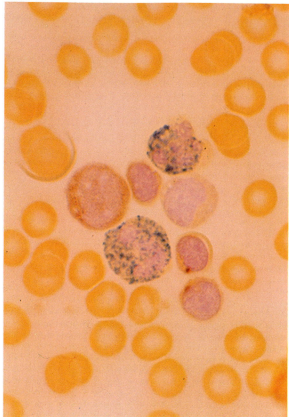

381

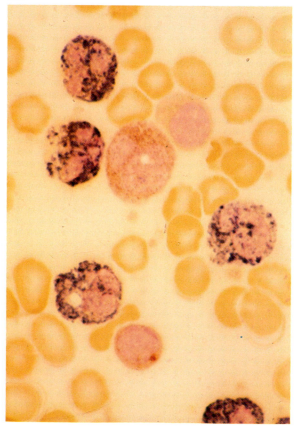

383

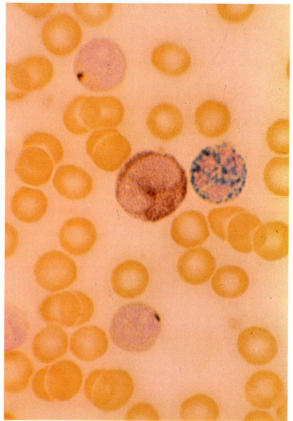

384–395. *Reactive 'immunoblasts' and 'virocytes' arising in various infective states. These activated lymphocytes are generally PAS, acid and alkaline phosphatase and esterase negative.*

384. Blood smear from a patient with infectious mononucleosis. Apart from a central monocyte (with phagocytic vacuoles) and a segmented neutrophil, the nucleated cells are all lymphocytes. The mitotic figure (in telophase) shows the increased cytoplasmic basophilia often seen in the abnormal lymphocytes of this disease. Mitoses in mononuclear cells are very uncommon in normal peripheral blood, but much more often seen in activated lymphocytes.

385 and 386. Further examples, from two different patients, of the abnormal morphology of lymphocytes in infectious mononucleosis. The nuclear chromatin in the most abnormal cells is still too coarse and the cytoplasm too abundant for real confusion to exist between these cells and lymphoblasts.

385

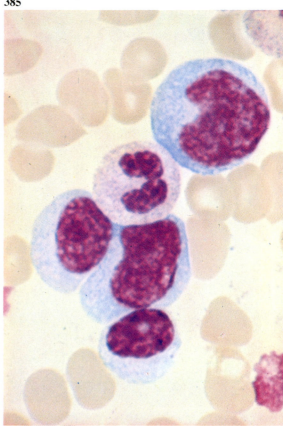

384

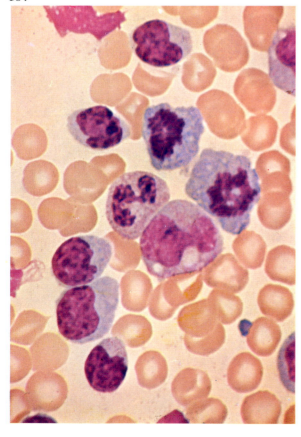

386

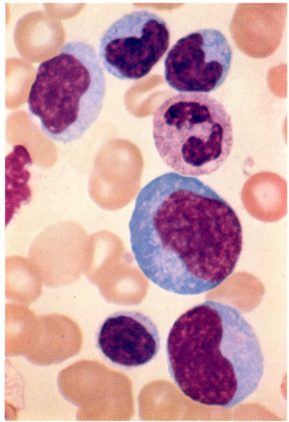

387. An immunoblast in the marrow of a patient with ALL in remission, but with a viral infection. Such cells must not be confused with leukaemic blast cells.

388. A disrupted RE cell and a binucleated immunoblast from the same case.

389. Three stabs, two lymphocytes and five immunoblasts from an infective episode during remission emergence in AML. One of the immunoblasts shows early 'flaring' of the cytoplasm (cf. **472–474**).

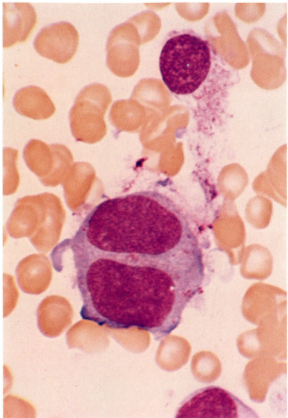

388

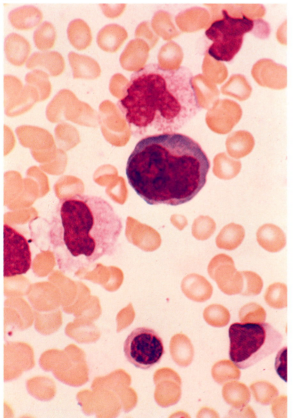

387

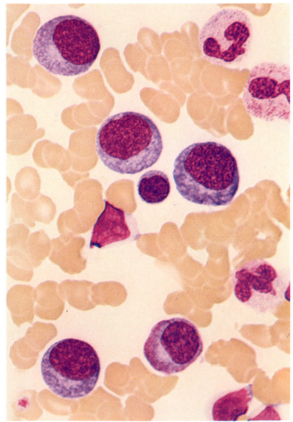

389

390. Two neutrophils, two lymphocytes, a monocyte and four immunoblasts (activated lymphocytes) from the buffy coat of a leukaemic patient in remission but with a mycoplasma infection. These cells must be distinguished from early monocytes or leukaemic blast cells.

391. PAS reaction, showing two stab cells, one lymphocyte and eight immunoblasts from the same buffy coat as in **390**. The immunoblasts are almost entirely PAS negative.

392. Peroxidase reaction on buffy coat from a patient with hairy cell leukaemia (HCL) (one hairy cell at top right) showing 24 per cent of immune reactive plasma cells resulting from acute bacterial infection. The neutrophils show normal peroxidase positivity, while the remaining cells are negative.

391

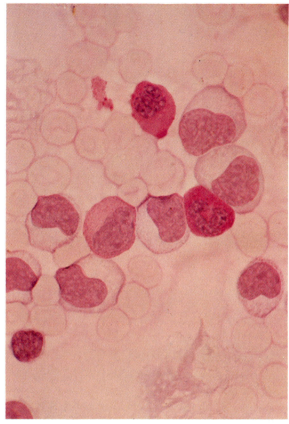

390

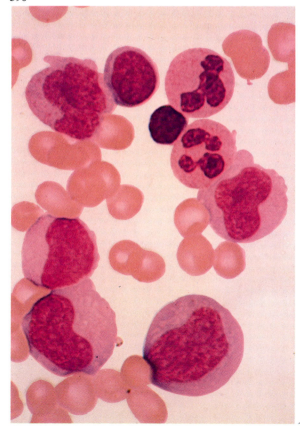

392

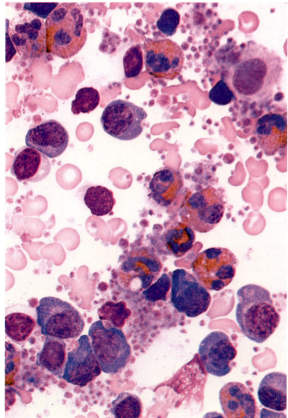

393. A vacuolated immunoblast and two stab cells with toxic granules, more conspicuous in one of them, from the blood of a child with a pseudomonas infection.

394. A binucleated immunoblast, showing tendency to nuclear distortion and peripheral cytoplasmic basophilia, from a patient with a viral infection.

395. A mononuclear immunoblast with twisted nucleus and fine azurophil stippling of cytoplasm, but with basophilia at the periphery, from the same case.

394

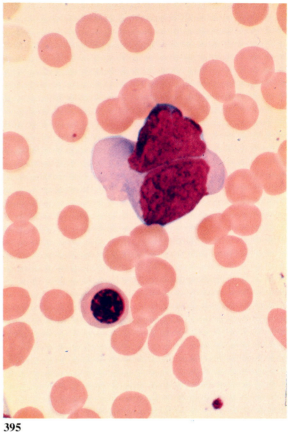

393

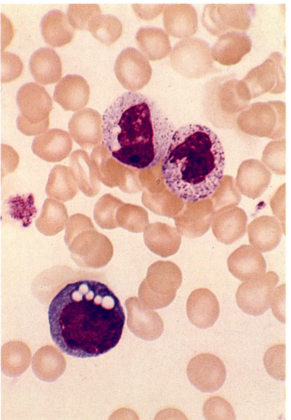

395

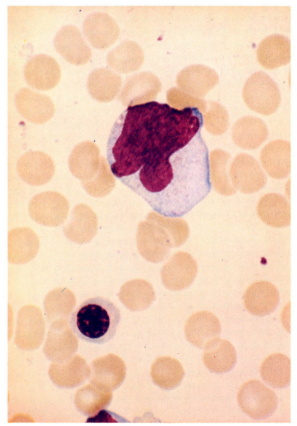

396–407. *Prolymphocytic leukaemia (PLL).*

396. PLL, Romanowsky stain: the cells show more basophilic cytoplasm and more pronounced nucleoli than do lymphocytes but have moderately pachychromatic nuclei, unlike blast cells.

397. PLL, Romanowsky stain, a T-cell case: the large cell size and moderately pachychromatic nuclear pattern, with occasional variable nucleoli, contrasts sharply with the two normal lymphocytes present.

398. PLL, Leishman stain on another case; this was of B cell origin. The nuclear cleavage and invaginations sometimes seen in PLL cells are well shown in this case.

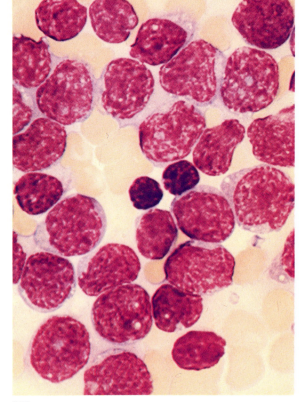

397

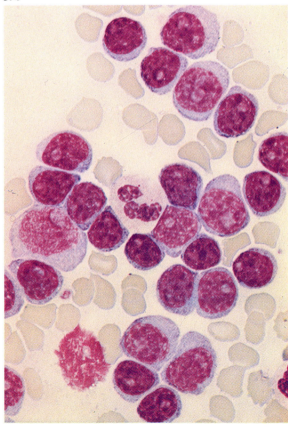

396

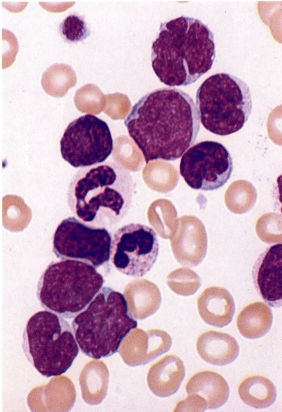

398

399. SB on PLL: a normally reacting eosinophil with prolymphocytes of variable cytology but uniformly negative reactions.

400. PLL, PAS reaction: the prolymphocytes are essentially PAS negative in this T-cell case.

401. PAS reaction on a B-cell PLL case. The prolymphocytes are again essentially negative but one, more immature, blast cell (top right) shows a rim of granular PAS positivity.

400

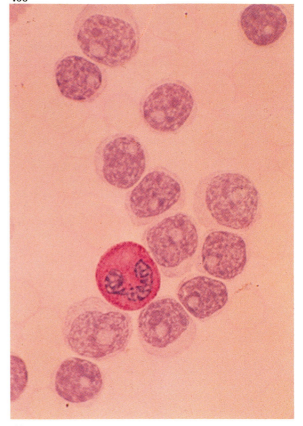

399

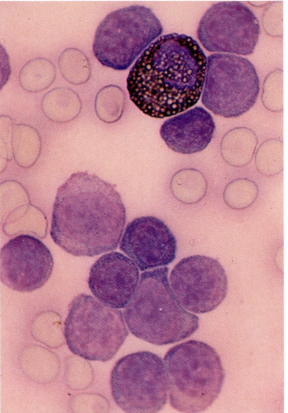

401

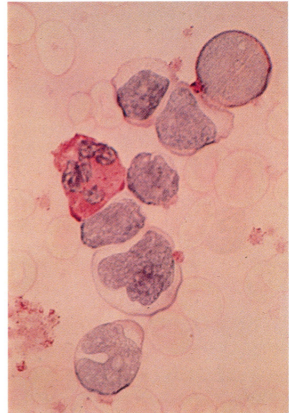

402. PLL: acid phosphatase reaction. In this T-cell case all the prolymphocytes show strong localised reaction.

403. PLL: dual esterase on the same T-cell case as shown in **397, 400** and **402**. The reaction is almost negative.

404. PLL: acid phosphatase reaction. Another T-cell case showing variable positivity, about half the cells having one or two coarse granules, as often seen in mature Tμ cells.

405. Dual esterase reaction in the same case as **404**; about half the prolymphocytes showed moderately strong localised BE positivity.

406. Acid phosphatase reaction in a B-cell PLL. The field shows a normally positive polymorph and seven prolymphocytes with occasional granules of coarse positivity scattered mostly over the nucleus rather than localised in the paranuclear zone.

407. Dual esterase on the same case as in **406**. Two CE positive polymorphs and a BE positive normal Tμ cell contrast with the virtually negative B prolymphocytes.

404
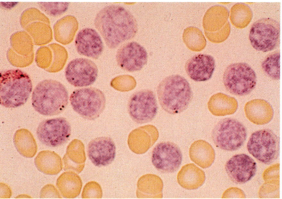

405
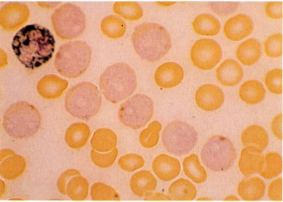

402
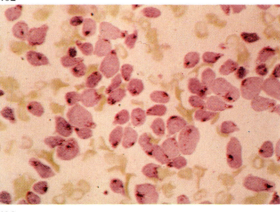

403
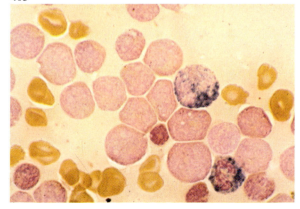

406
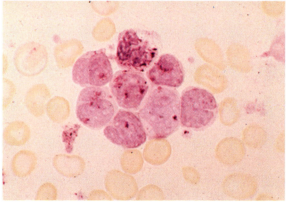

407
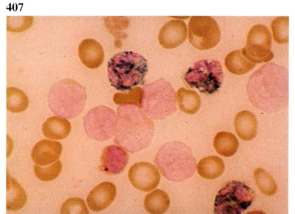

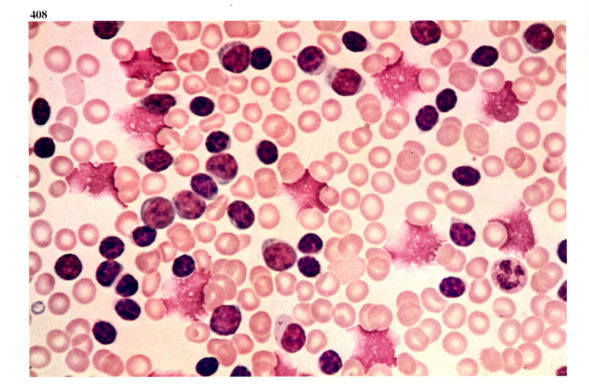

409

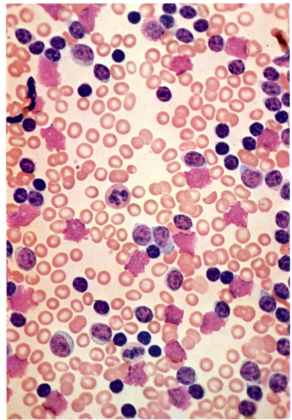

408. A general view of a peripheral blood smear from chronic lymphocytic leukaemia (CLL). The intact cells are nearly all lymphocytes, showing some variation in morphology probably in parallel with maturity. The smeared, disrupted cells are typically numerous in this disease. The spread nuclear remnants are sometimes known as 'Gumprecht's shadows'.

409. Another example of the blood picture in CLL. The range of morphological variation is not great, but a few precursors – prolymphocytes or lymphoblasts – can be seen, and the smeared cells are again conspicuous.

410. A high power view of the peripheral blood in CLL. Again an occasional precursor is seen.

411. PAS reaction in CLL. The lymphocytes show a pattern of positivity like that of normal lymphocytes, but a higher proportion of cells shows positivity than in a normal specimen.

412. Another example of PAS reaction in CLL.

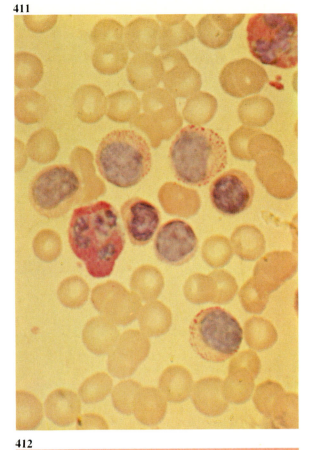

411

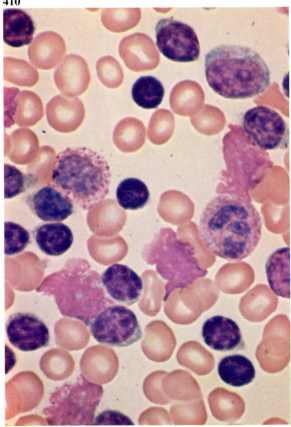

410

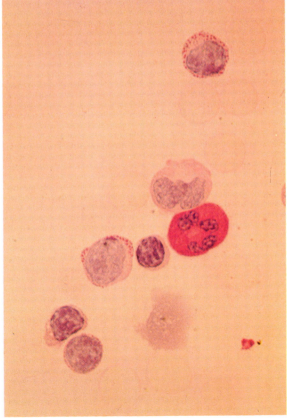

412

413. Acid phosphatase in CLL. Most cells show scattered positivity without marked paranuclear concentration.

414. Dual esterase in CLL. The lymphocytes show mostly negative reactions; a few have scattered weak BE positivity. Normal CE reaction in a polymorph and BE positivity in a monocyte are shown.

415. Dual esterase in another case of CLL, showing an occasional crescentic positive reaction, similar to that usually seen in hairy cell leukaemia.

414

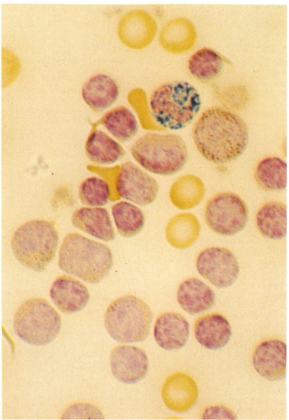

413

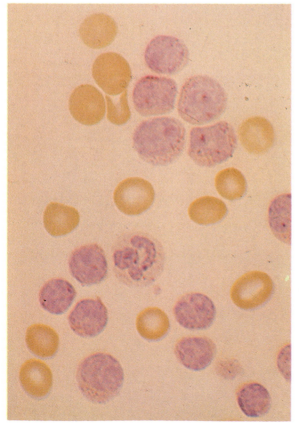

415

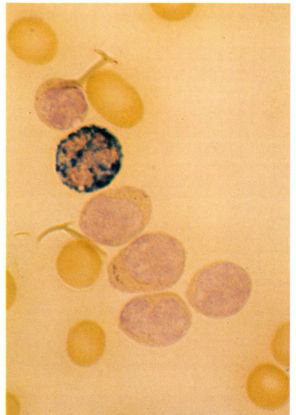

416. Four lymphoma cells from the blood of a patient with longstanding non-Hodgkin's lymphoma recently spreading to involve bone marrow and blood. The cells resemble neither mature lymphocytes nor leukaemic lymphoblasts, but have intermediate characteristics. One cell shows a tendency to nuclear cleaving. These cells are of 'centrocytic' or 'follicular centre cell' (FCC) type.

417. Another example of lymphoma cells in the blood: this represents a leukaemic phase of a secondary centroblastic lymphoma. There are multiple small nucleoli visible in the irregular nuclei and cytoplasmic basophilia is moderate.

418. A third example of lymphoma cells in the blood – in this case the cells have deep cytoplasmic basophilia and multiple nucleoli and are of immunoblastic character.

417

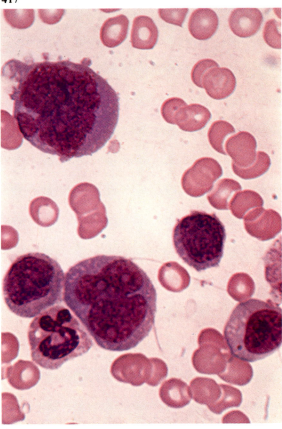

416

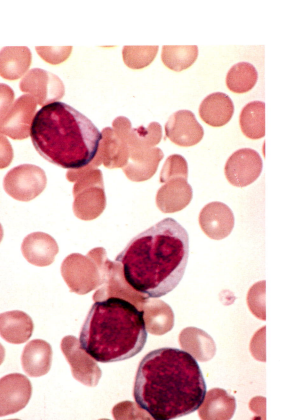

418

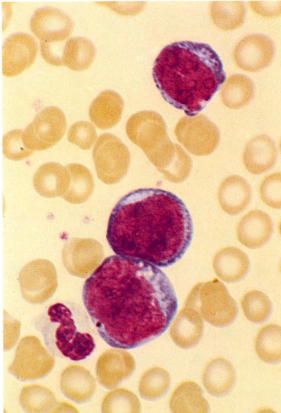

419. Lymphoma cell from the specimen illustrated in **416**, stained with the PAS reaction, to show coarse granules and two small blocks of positivity. A polymorph reacts normally. Centrocytes of this kind more often show PAS positivity when circulating in the leukaemic phase than do similar cells in lymph node imprints.

420. Strong PAS reaction, with rings of moderately coarse granules around the nuclei in circulating lymphoma cells, from the case illustrated in **417**. Fine and moderately coarse granular positivity is present.

421. Almost negative PAS reaction in the circulating neoplastic immunoblasts from the case illustrated in **418**.

420
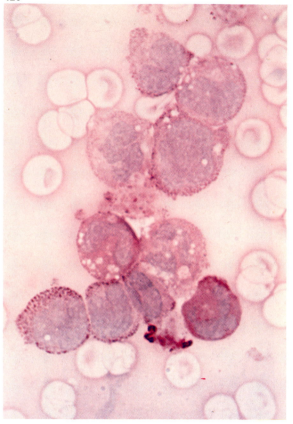

419
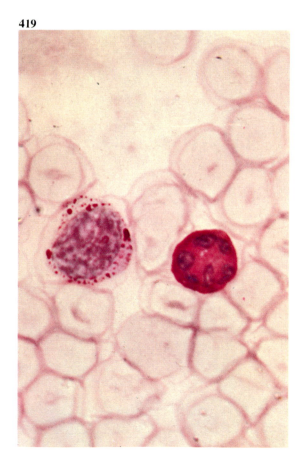

421
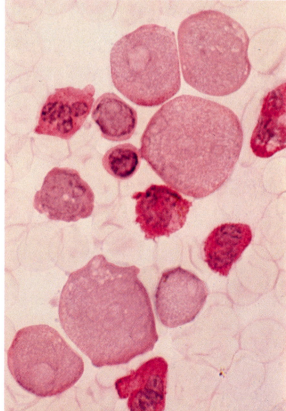

422. Acid phosphatase reaction in the neoplastic immunoblasts – a few scattered granules only are seen in the cytoplasmic rim.

423. Dual esterase reaction on the same preparation, showing three CE positive neutrophils and three malignant immunoblasts having BE positivity partly scattered and partly localised.

424. Scattered BE positivity in three neoplastic immunoblasts, a normally BE positive monocyte, a T-lymphocyte with localised 'dot' BE positivity and a CE positive neutrophil.

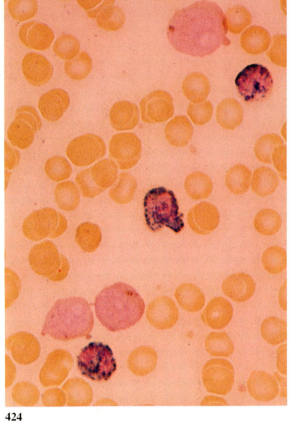

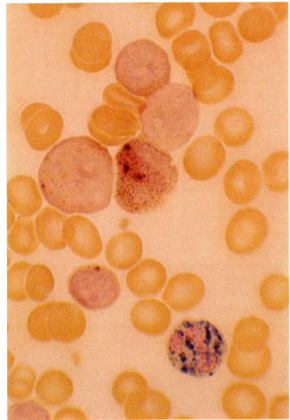

425–428. *Blood smears at stages during the progression of a lymphoblastic lymphoma, convoluted (T-cell) type, to a picture virtually indistinguishable from acute lymphoblastic leukaemia.*

425. A lymphocyte and two lymphoma cells at a stage when involvement of the blood first occurred, with small numbers of cells.

426. The lymphomatous lymphoblasts now predominate and the total leucocyte count has risen above normal.

427. A fully leukaemic picture.

426

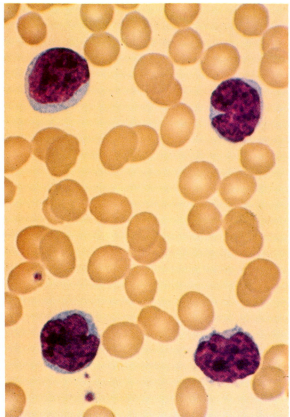

425

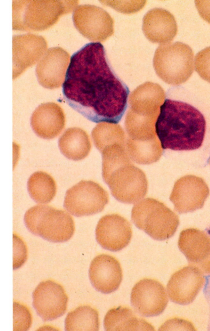

427

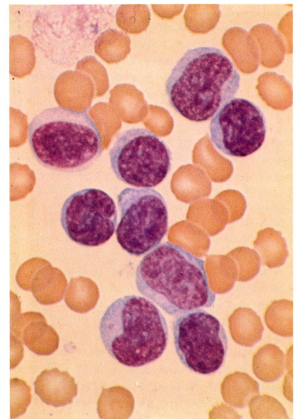

428. The cells show coarse PAS positivity of the lympho-blastic pattern.

429. Acid phosphatase reaction in these cells shows strong localised positivity in most cells.

430. Dual esterase reaction in the same cells shows negativity to BE but a few scattered CE granules only, in the T lymphoblasts.

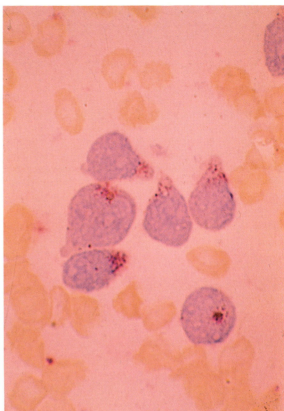

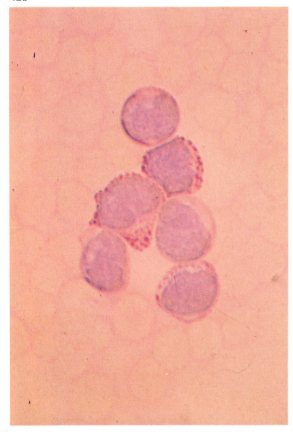

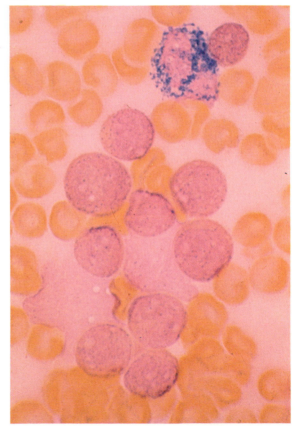

431–433. *Circulating lymphoma cells from a case of follicular lymphoma (mixed small and large FCC tumour, centroblastic/centrocytic type).*

431. Romanowsky stain: a monocyte, two neutrophils and seven lymphoid cells of variable cytology, including a nucleolated centroblast.

432. PAS reaction: variable positivity in centroblasts and centrocytes.

433. Alkaline phosphatase reaction showing positivity in lymphoma cells – a most unusual finding in the peripheral blood. The seven neutrophils present show positivity ranging from + to +++.

432

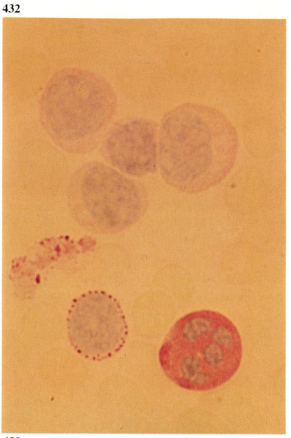

431

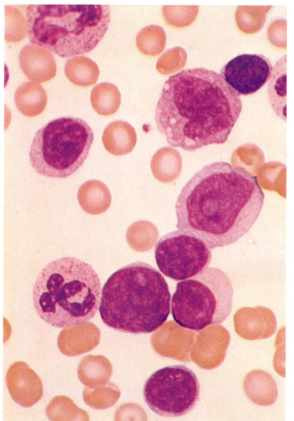

433

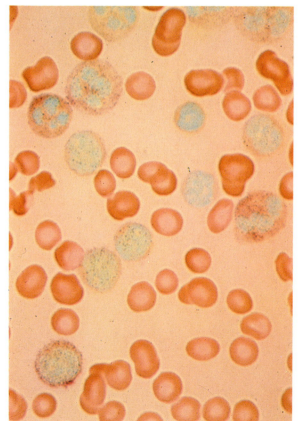

434. A circulating lymphoma cell from a case of lympho-blastic lymphoma, B-cell type, with only occasional lymphoma cells in the peripheral blood. The field also contains two lymphocytes and a segmented neutrophil.

435. Marrow infiltration in the same case: the lymphoma cells show a more lymphoblastic appearance, with higher nuclear-cytoplasmic ratio, than those in the peripheral blood.

436. Chromosome spreads from the blood, made after 70 hours in culture, showed numerous abnormal meta-phases with 49 chromosomes, as illustrated here, including a marker large acrocentric chromosome, here shown encircled.

435

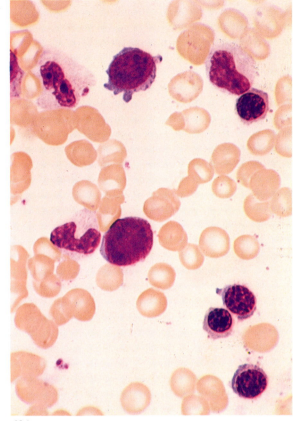

434

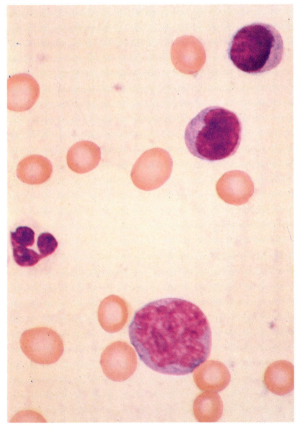

436

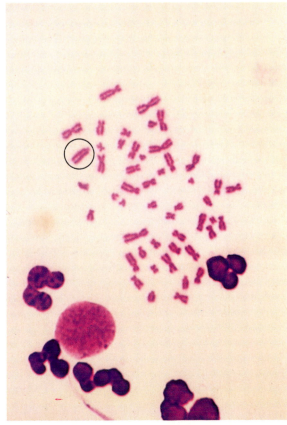

437–440. The first two figures are of Romanowsky stains and the latter two of Sudan black and PAS preparations respectively, from the bone marrow of a child with Burkitt's lymphoma. An appearance comparable with lymphomatous or acute leukaemic invasion is seen. This type of extensive marrow replacement by tumour is unusual in Burkitt's lymphoma.

The morphology of the cells is unlike that of typical leukaemic lymphoblasts; the nuclear chromatin is coarser and more stranded and most cells are PAS negative. The vacuolation, characteristic of the tumour cells in this disease, is well shown, but similar vacuoles may sometimes be seen in acute leukaemias and this feature is not pathognomonic.

438

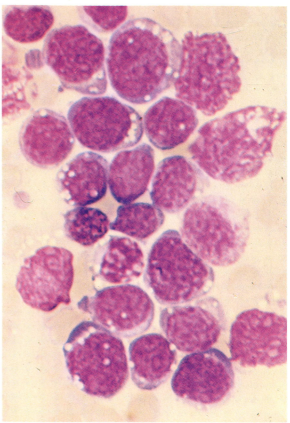

437

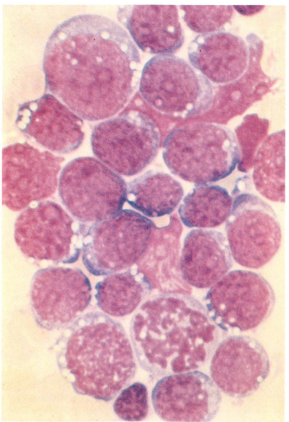

439

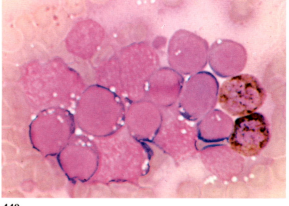

440

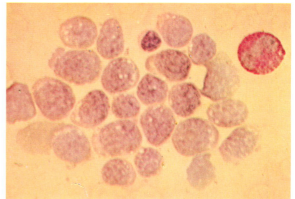

441–454. *Preparations from typical hairy cell leukaemia (HCL) – leukaemic reticuloendotheliosis (LRE). In all these smears monocytes are conspicuous by their absence.*

441. HCL. Leishman stain of peripheral blood. Two neutrophils, a normal lymphocyte and four hairy cells (HCs) showing the typical eccentric nucleus with moderately coarse chromatin, the pale slate-blue cytoplasm and the fine surface projections or hairs. A rod-shaped negatively-staining inclusion in one of the HCs may represent a ribosome-lamella (R-L) complex.

442. HCL. Leishman stain of peripheral blood from another case. A group of six HCs together with one lymphocyte and two segmented neutrophils. The nuclear staining of the HCs shows the sponge-like or checkerboard pattern often seen.

443. HCL. Leishman stain of another example, with an HC showing the conspicuous cytoplasmic vacuolation which is often a feature of these cells, together with a normal lymphocyte, a neutrophil polymorph and an immunocyte or plasma cell, from a patient with HCL and a virus infection.

442

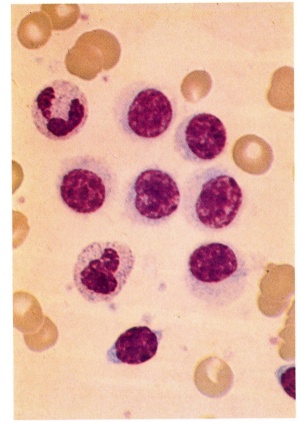

441

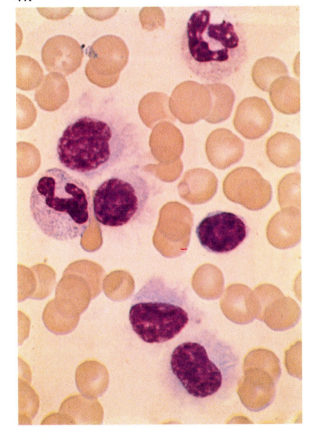

443

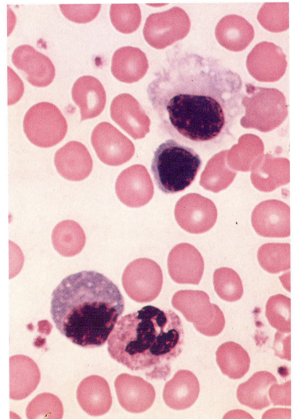

444. HCL. Leishman stain of a bone marrow smear showing a cluster of HCs with the azurophilic inclusions sometimes present and especially seen after short term culture.

445. Consecutive PAS stain on the same field as in **444**. The tinge and fine granular positivity is typical of HCs.

446 and 447. HCs in peripheral blood before and after a period of 48 hours in culture, showing increase in cytoplasmic granularity and the occurrence of mitosis.

448. A further example of HCs after 48 hours in culture, showing phagocytosis of red cells and bacteria.

446

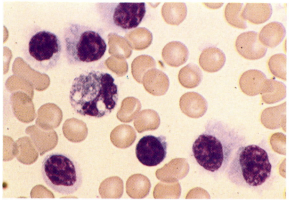

447

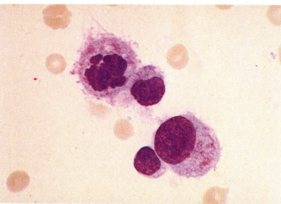

444

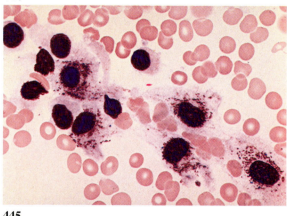

445

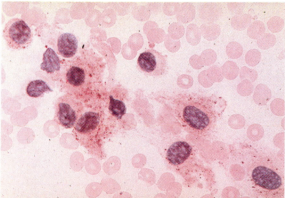

448

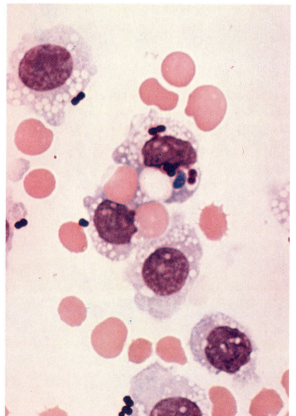

449. HCL. Sudan black reaction, showing normal positivity in two neutrophils, a negative lymphocyte and two negative HCs.

450. HCL. PAS reaction, showing typical diffuse and granular positivity in HCs. A normally reacting neutrophil and a negative lymphocyte complete the field.

451. HCL. Acid phosphatase reaction. Five polymorphs, four probable Tμ lymphocytes and two HCs all showing typical positivity. That in the HCs is tartrate resistant.

450

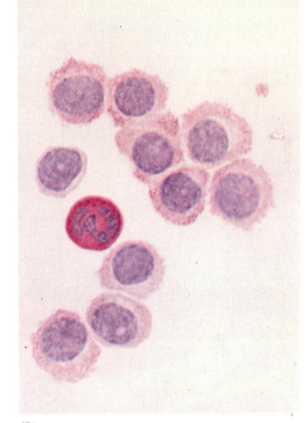

449

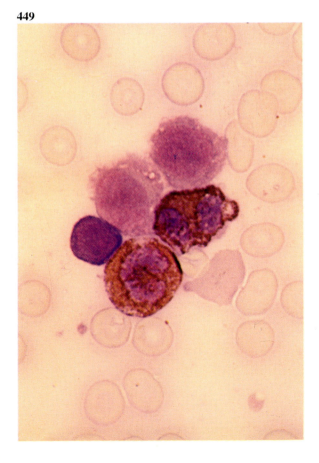

451

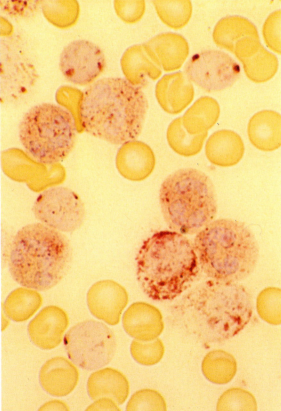

452. A further example of acid phosphatase staining in HCL, showing a neutrophil, two Tµ cells and five HCs, several containing rod-like structures with positive staining, possibly representing R-L complexes. Such inclusions are seen more commonly after acid phosphatase staining than in Romanowsky preparations.

453. Dual esterase reaction in HCL. Two CE positive neutrophils, a Tµ cell with localised BE positivity and an HC with granular positivity tending to show a crescentic localisation at either side of the nucleus.

454. Alkaline phosphatase reaction in neutrophils in HCL. The score is almost invariably high as in this case where most neutrophils show +++ to ++++ positivity.

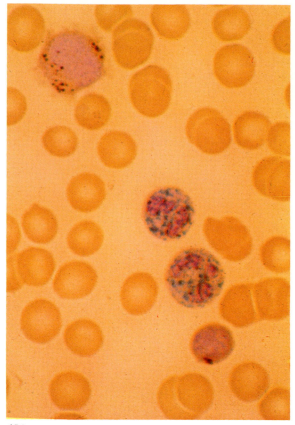

453

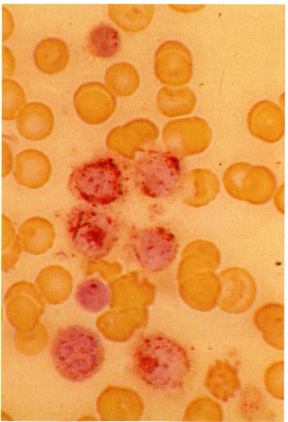

452

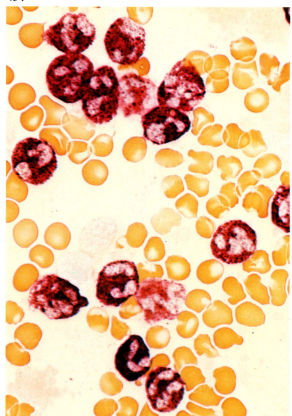

454

455–459. *An example of atypical HCL, with high leuco-cytosis, presence of monocytes, normal LAP score and chronic course.*

455. Romanowsky stain, showing more central nuclei, more conspicuous nucleoli, and coarser cytoplasmic projections than usual in HCs.

456. Sudan black stain, showing three negative HCs, two positive neutrophils and two monocytes with variable scattered positive granules.

457. PAS stain shows the usual HC pattern, quite unlike that of, for example, CLL or PLL.

458. Acid phosphatase shows a more markedly polar distribution than in typical HCL.

459. Dual esterase in atypical HCL. Two CE positive neutrophils, one heavily vacuolated, two BE positive vacuolated monocytes and three nucleolated HCs, one showing polar BE positivity.

456

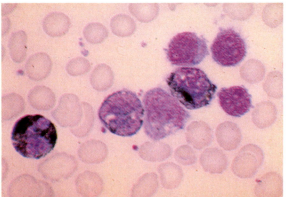

457

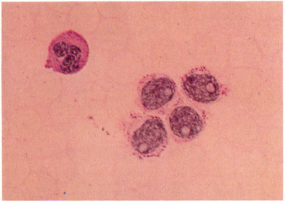

455

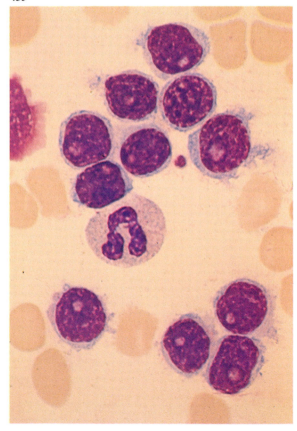

458

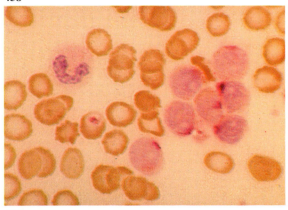

459

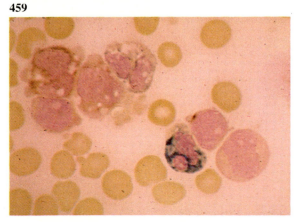

460–462. *Circulating abnormal neoplastic T cells from a case of Sézary's syndrome.*

460. Romanowsky stain: the conspicuous nuclear convolutions shown on electron microscopy (**461**) of the same cells are barely detectable as nuclear creasing by light microscopy.

461. EM preparation from the same case. The circulating Sézary cells show a 'cerebriform' nuclear section; with striking invaginations. Glycogen granules are conspicuous in the cytoplasm.

462. PAS reaction: most cells in this case are negative, but occasionally fine or moderately coarse granular positivity is present. Nucleoli are well shown in some of the neoplastic cells, as is the creased or cleft appearance of some nuclei. The PAS reaction in Sézary cells is quite variable, some cases having much more marked granular positivity.

461

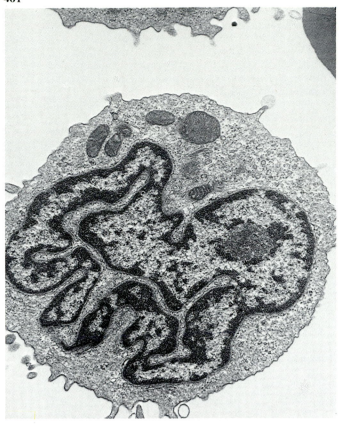

460

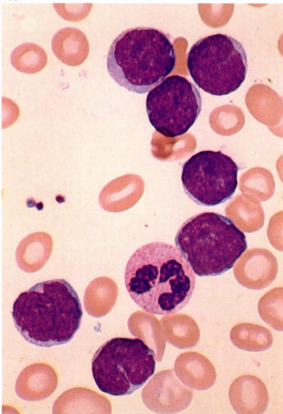

462

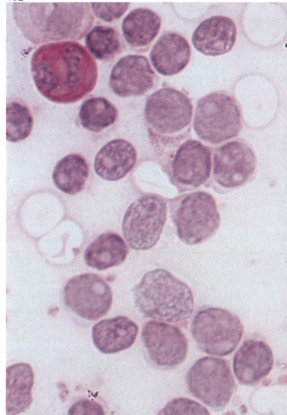

463. A nest of mature plasma cells from the marrow of a patient with cat-scratch disease, showing some reactive plasmacytosis. The vacuolation and tendency to budding and fragmentation of the cytoplasm may indicate enhanced activity.

464. A group of plasma cells with more immature features than those in **463**, from a patient with multiple myeloma.

465. Variability in the size, maturity and general staining characteristics of plasma cells in the marrow of a patient with multiple myeloma.

464

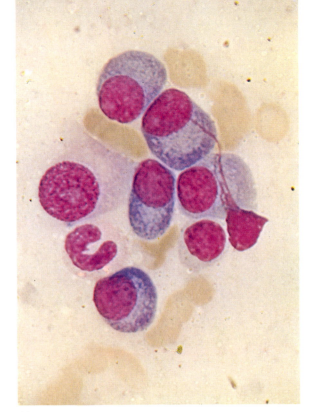

463

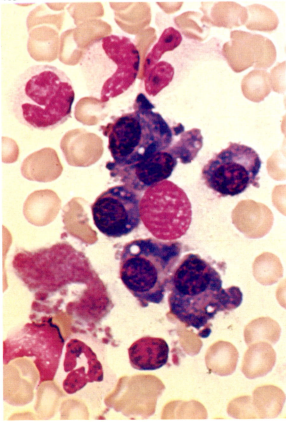

465

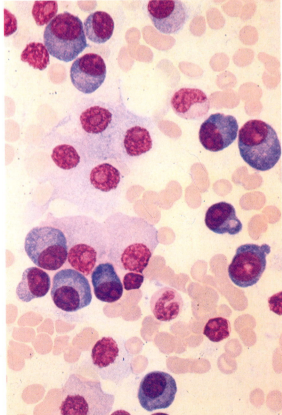

466. Circulating plasma cells from a case of plasma cell leukaemia with predominantly mature, non-nucleolated, cells. Note absence of rouleaux formation.

467. Acid phosphatase reaction in the same case: there is strong localised positivity but considerably less intense than usually seen in typical myeloma cells of bone marrow (see **491**, **492**).

468. Dual esterase reaction: the same case shows only very weak BE positivity in the plasma cells in comparison with the stronger reaction for BE or CE or both enzymes common in typical myeloma cells of bone marrow (see **493**).

469. A remarkable example of internuclear bridging in a case of myeloma. Most myeloma cells had 2 or 3 linked nuclei.

467
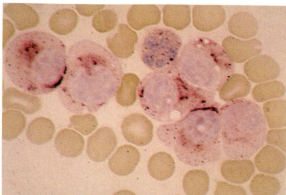

468
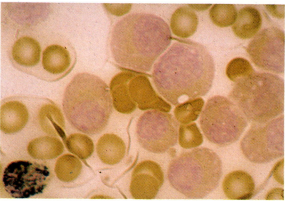

466
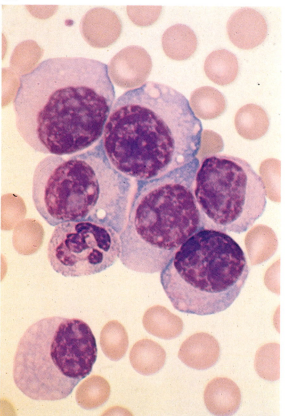

469
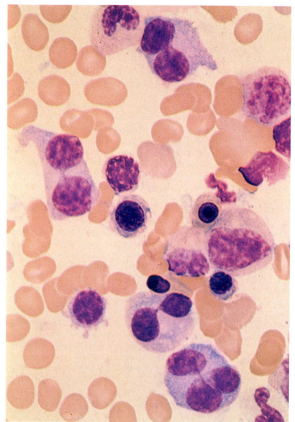

470. Plasma cells in myeloma, showing a remarkable tendency to cytoplasmic disruption.

471. A multinucleated plasma cell in myeloma.

472 and 473. The development of flaming cells. The smooth eosinophil component which makes up the 'flaming' character in certain plasma cells is usually first seen at the cell periphery. It often contrasts sharply with intense basophilia in the remaining cytoplasm.

471

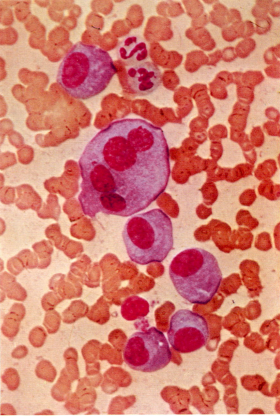

470

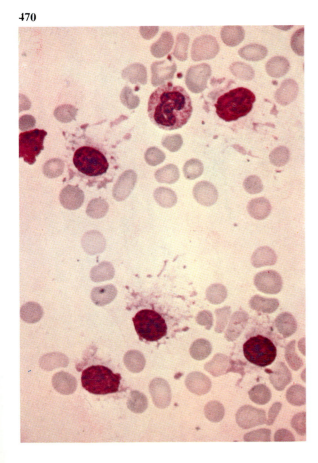

472

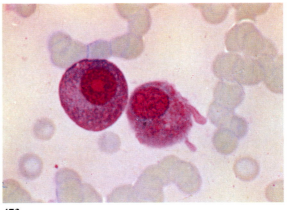

473

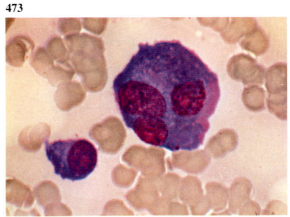

474. A pair of fully developed flaming cells.

475 and 476. The development of thesaurocytes – large plasma cells with dark, somewhat pycnotic nuclei and extensive fibrillary cytoplasm, sometimes having the appearance of division into compartments. These 'storage cells' usually have 'flaming' characteristics in their remaining cytoplasm.

477 and 478. Large nuclear inclusions, PAS positive, occasionally seen in plasma cells in myeloma (and in other conditions). Their significance is unknown; they are less likely than cytoplasmic inclusions to represent secretion or synthesis products.

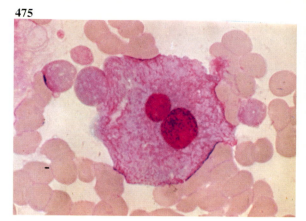

475

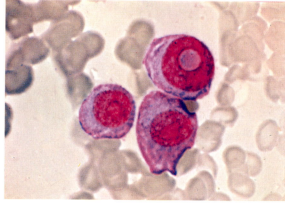

476

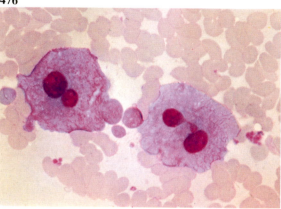

477

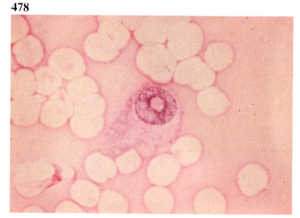

478

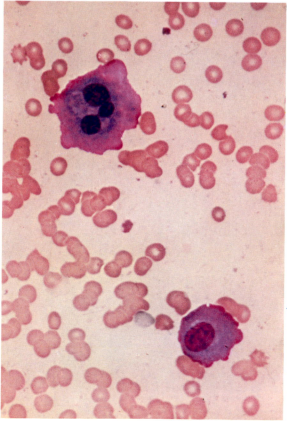

474

479 and 480. Plasma cells in myeloma, showing accumulation of spherical inclusions, bluish in colour, probably representing an abnormal concentration of immunoglobulin precursor. Cells may become full of these bodies (called by some authorities Russell bodies) and are then sometimes referred to as Mott cells.

481 and 482. Another type of spherical inclusion, eosinophil staining and PAS positive, found much less often in smears of plasma cells. This type of inclusion is also given the name Russell body, perhaps with more historical justification.

483. An unusual form of locular degeneration in the cytoplasm of a plasma cell from myeloma.

484. Azurophilic rods (resembling Auer rods, but negative to Sudan black, peroxidase and PAS staining) are not rare in plasma cells, but in this unusual myelomatous case the majority of the plasma cells contained many such rods.

481

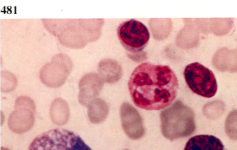

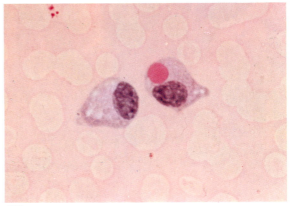

482

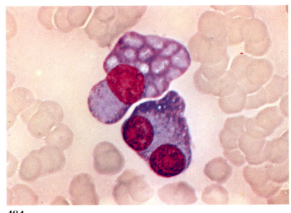

479

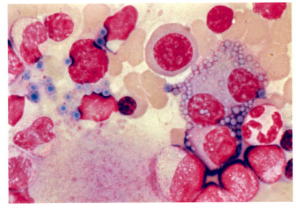

483

480

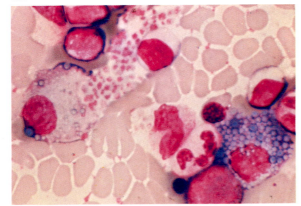

484

485. Crystalline inclusions of immunoglobulin are seen in most myeloma cells in this case. They appear as negatively stained against the basophilic cytoplasm.

486 and 487. An even more unusual case of myeloma, where nearly all the plasma cells were distorted by large crystalline inclusions, having some resemblance to Charcot-Leyden crystals, and presumably representing the product of disordered synthetic activity in the cells. The cell cytoplasm showed weak PAS positivity, but the inclusions were negative (**487**).

488–490. PAS reactions in plasma cells from cases of multiple myeloma. Reactions range from negative to weakly positive, with occasional granules against a weaker diffuse background of faint positive tinge. The spherical bluish inclusions illustrated in **479** and **480** are PAS negative.

487

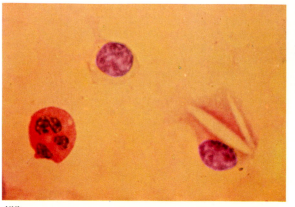

488

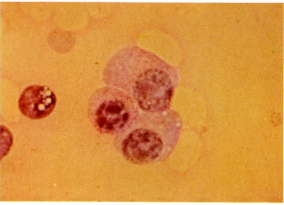

485

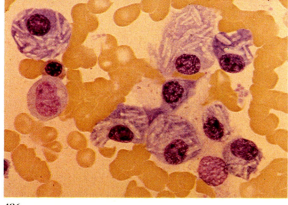

486

489

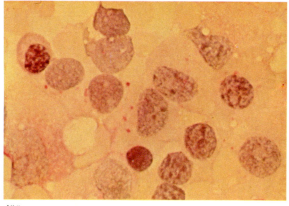

490

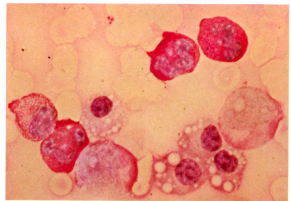

491. The one striking cytochemical characteristic of plasma cells, normal and pathological, is their consistently strong acid phosphatase reaction, illustrated in this general view of a bone marrow smear from a patient with myeloma.

492. A second example of acid phosphatase reaction in myeloma cells. The more sharply particulate deposit arises from the use of naphthyl AS-BI phosphoric acid as substrate, compared with α-naphthyl acid phosphate, as used for **491**.

493. Dual esterase reaction in a marrow smear from a patient with myeloma. The cells contain scattered CE positivity in this case, but no BE as is sometimes found.

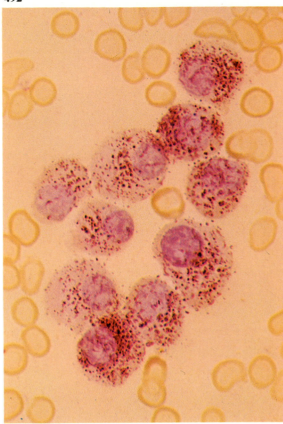

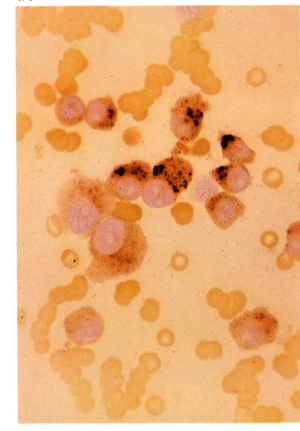

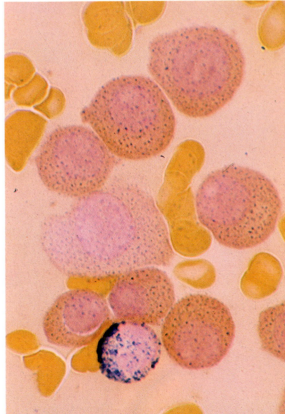

494–496. *Transformation of myeloma to AML.*

494. Myeloma marrow with chiefly mature well differentiated myeloma cells but with occasional blast cells of myelomonocytic cytology.

495. A higher power view to contrast the range of myeloma cell morphology with two nucleolated myeloid blast cells.

496. Marrow smear from the same patient at a later stage of transformation, showing predominance of AMML blast cells with scanty residual plasma cells and rare later granulocyte stages.

495

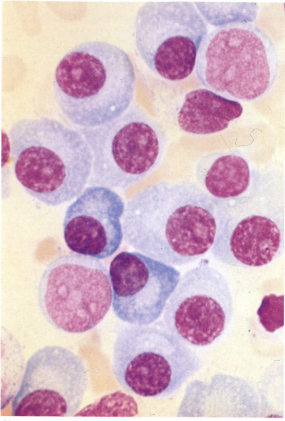

494

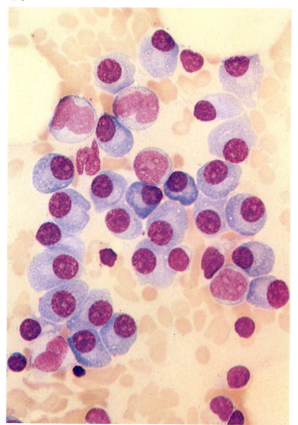

496

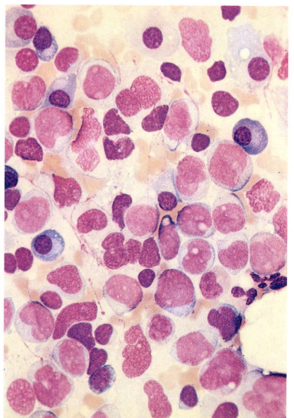

497–500. *The marrow picture in Waldenstrom's macro-globulinaemia is not specific, but appearances such as these are very suggestive and would indicate the need for investigation of the immunoglobulin pattern.*

497. A bone marrow smear from a patient with macro-globulinaemia. Lymphocytes, mostly with disrupting or minimal cytoplasm, predominate. There is also a tissue mast cell present.

498. Another example of bone marrow cytology in macroglobulinaemia. The red cells show conspicuous rouleaux formation, the lymphocytes predominate and have scanty cytoplasm, and there is a plasma cell present as well as a very densely granular tissue mast cell. An alternative name for this last cell, 'basophil ball cell', would here be descriptively apt.

499. Another field from the same preparation, with similar lymphocytes but a disrupted tissue mast cell.

500. The neoplastic cells in this field from the bone marrow of another case show occasional 'lympho-plasmacytoid' features. Phagocytic RE cells are also conspicuous.

498

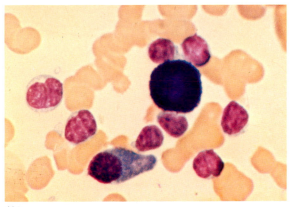

499

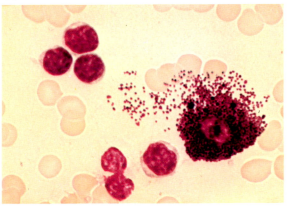

497

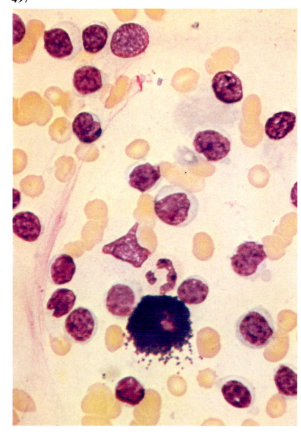

500

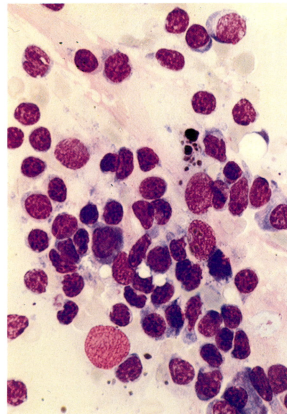

501. PAS stain in macroglobulinaemia: there is scattered granular PAS positivity in the cytoplasm of two RE cells but the lymphocytes and plasmacytoid cells in this disease show little or no reaction.

502. Dual esterase stain in macroglobulinaemia showing several tissue mast cells with coarse CE positivity in a marrow fleck. The neoplastic lymphoid cells and RE cells in the fleck show only weak scattered granular positivity to both BE and CE.

503. Free-iron stain on bone marrow smear in macroglobulinaemia. The excessive amount of free iron, both scattered and in macrophages, is evident.

504. Romanowsky stain of marrow from macroglobulinaemia, showing a macrophage heavily laden with iron.

505. A similar cell to that in **504**, stained with the Prussian blue stain for free iron.

503

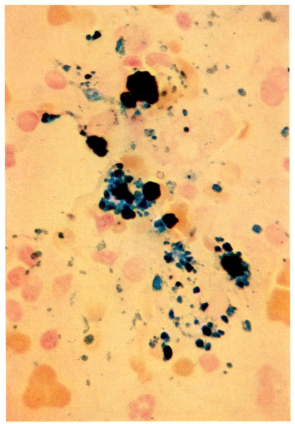

501

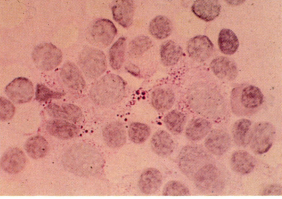

502

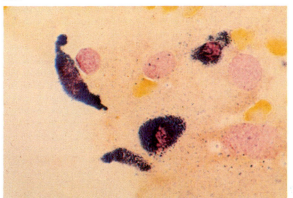

504

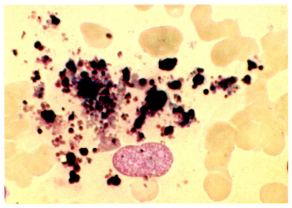

505

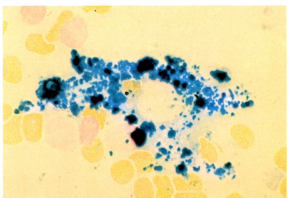

Part 4
Miscellaneous cells from bone marrow or blood smears, reticulo-endothelial cells, osteoclasts and osteoblasts, foreign cells and parasites

Reticulo-endothelial cells (reticulum cells, histiocytes, macrophages) are common in the bone marrow and have been illustrated several times previously. They take up foreign particles, free iron, fat globules, specific granules from disrupted granulocytes, and other cell fragments, and therefore often contain phagocytosed material. Their cytoplasm appears fragile and is readily broken up in the smearing process or stretched out between neighbouring cells so that the cytoplasmic outlines may be difficult to recognise. They may contain various inclusions – sea-blue histiocyte material, pseudo-Gaucher cell birefringent lipid or blue crystals and grey-green crystals. In certain lipid storage diseases they appear grossly swollen with abnormal fibrillary or globular deposits of lipids. The Leishman-Donovan bodies, the protozoal parasites of Leishmaniasis, appear most prominently in reticulo-endothelial macrophages. Malignant histiocytosis is a rare neoplastic state of the RE cell and even rarer is histiocytic leukaemia. Examples of each are illustrated.

Apart from reticulo-endothelial cells, most of the other cells illustrated in this section are less commonly encountered, but they have highly characteristic cytological features, and once these features are appreciated the cells are unlikely to be confused with normal or abnormal variants of more common cell lines in the bone marrow or blood.

Several different examples of tumour cells invading the marrow are shown. Although an isolated tumour cell may rarely be identified as such in the absence of more typical cell clumps, the feature which allows identification in most instances is the occurrence of cell nests, frequently partially syncytial, of cells not belonging to any haemic series. The identification cannot often go beyond 'metastasising tumour cells'.

Certain of the more common parasites which may be seen in smears of blood or bone marrow are also illustrated. They are mostly found in tropical or subtropical zones but their appearance should be recognised by all haematologists, even though what may actually be seen under the microscope (as in the photographs here) does not always provide detail comparable with the diagrams and paintings in parasitology texts.

506. A phagocytic reticulo-endothelial (RE) cell containing various particles of cellular debris.

507. This is probably an RE cell with phagocytosed eosinophil granules from a disrupted myelocyte. The alternative possibility is that it represents a partially smeared degenerating eosinophil myelocyte.

508. A group of RE cells with poorly outlined filmy cytoplasm, from the bone marrow in a case of pernicious anaemia.

507

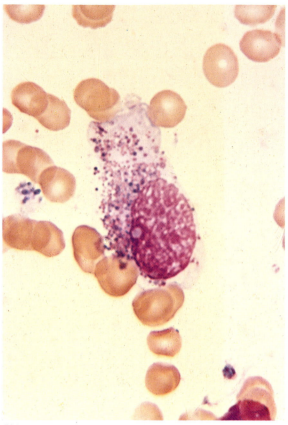

506

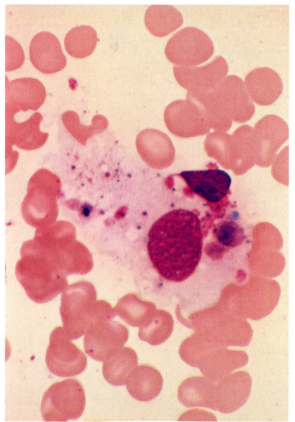

508

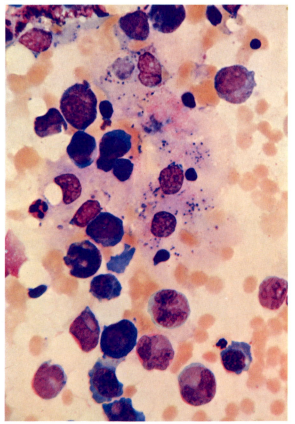

509. A group of three cells of the monocyte-macrophage system in buffy coat of peripheral blood from a patient with bacterial endocarditis. The middle cell is clearly a macrophage whereas the contiguous mononuclear cells are still of vacuolated monocytic cytology.

510. A similar macrophage in the buffy coat of this patient, stained to show free iron content.

511. Another example of the monocyte-macrophage sequence in peripheral blood.

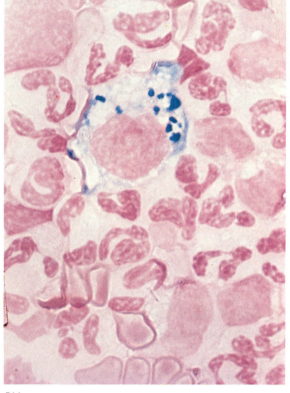

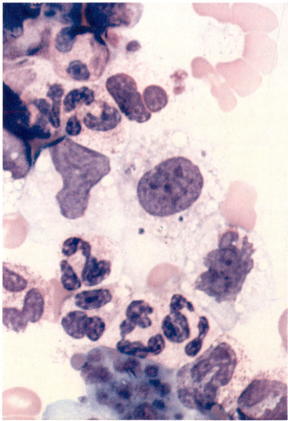

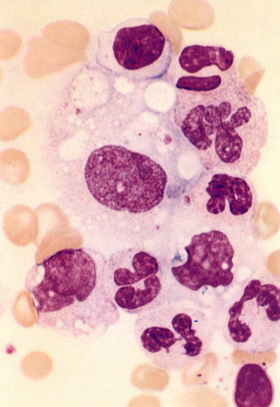

512–535. *Phagocytic RE cells with various cytochemical reactions and inclusions.*

512. Romanowsky stain: a multinucleated RE cell with remnants of several ingested cells, including a polymorph, and multiple granules, most probably iron.

513. Sudan black stain: strong positivity, probably surrounding phagocytosed material, including free iron particles. There are scattered SB positive granules in a monocyte.

514. PAS reaction: three RE cells, one with scattered PAS positive granules and the other two with weak PAS reaction but chiefly scattered free iron particles spreading across the field diagonally.

513

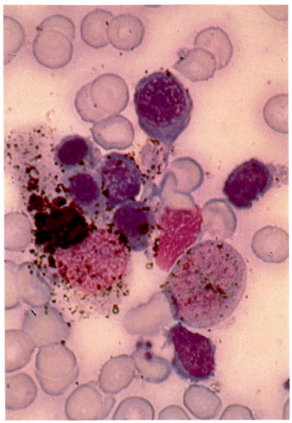

512

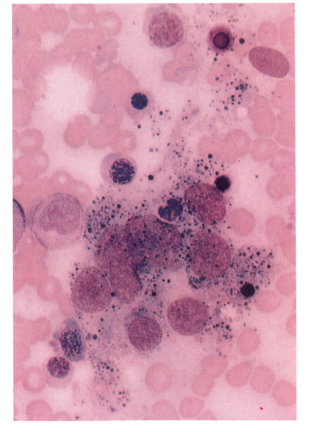

514

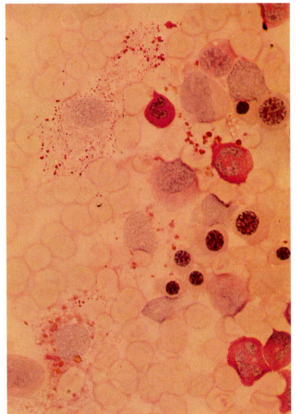

515. Alkaline phosphatase reaction, showing strong positivity in a marrow RE cell.

516. Acid phosphatase reaction: three strongly positive RE cells, containing phagocytosed cell remnants and large particles of free iron, heavily coated with acid phosphatase.

517. Dual esterase reaction: the phagocytic RE cell shows the typical very strongly positive BE reaction.

516

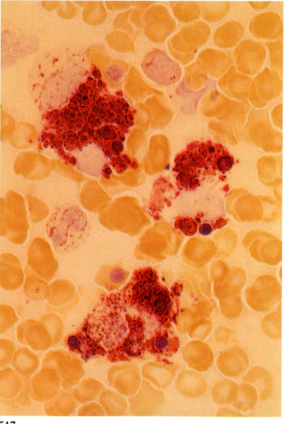

515

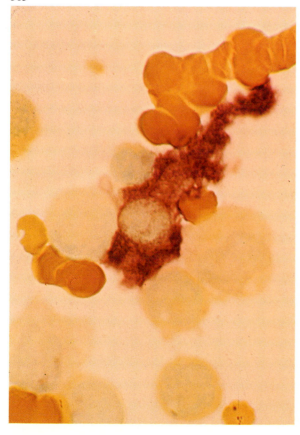

517

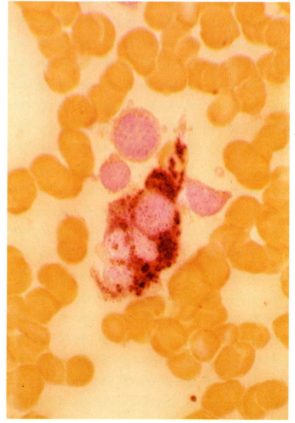

518. Faintly visible blue crystals together with coarser iron granules in a distended RE cell from the marrow in a case of CDA type I. RE cells containing such blue crystals, and pseudo-Gaucher cells, RE cells with crumpled swollen cytoplasm, are found not uncommonly in CDA.

519 and 520. Increased visibility and clear birefringence of these crystals, with reversal of refringence on 90° rotation when the same field as shown in **518** is looked at with polarised light.

521 and 522. Another Gaucher-like cell with duplicate under polarised light showing refringence. This specimen was from a case of CML, a condition in which the pseudo-Gaucher cell phenomenon is also not infrequent.

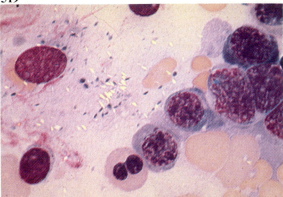

519

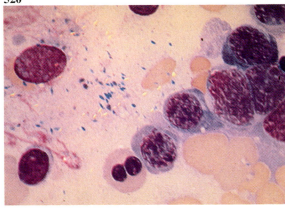

520

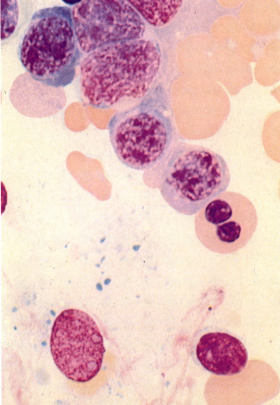

518

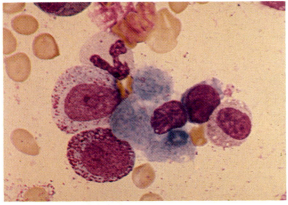

521

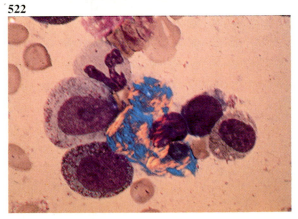

522

523. Sea-blue histiocyte (SBH): an RE cell, binucleated in this instance, containing a mass of blue granules in the Romanowsky stain. Such cells are seen especially in the benign genetic disorder of 'sea-blue histiocytosis', and secondarily in myeloid leukaemia and in dysmyelopoietic and dyserythropoietic states.

524 and 525. An SBH, consecutively stained by Leishman and free iron stains, to demonstrate that the granular material is chiefly negative for iron.

526 and 527. An iron-laden macrophage with some sea-blue material similarly stained consecutively for comparison. In this case much of the heavier granularity contains free iron.

524

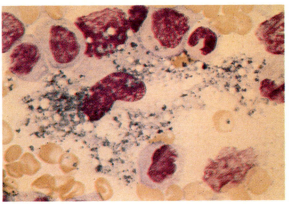

525

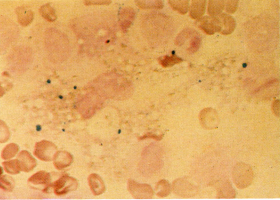

523

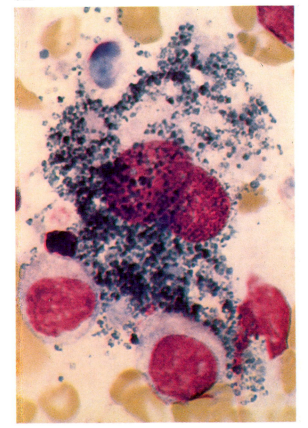

526

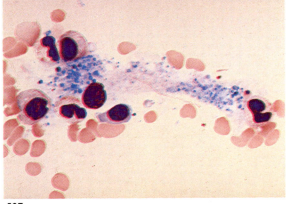

527

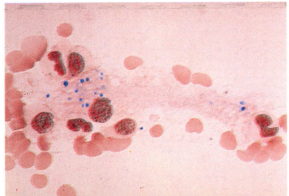

528. Grey-green crystals in a disrupted RE cell. These inclusions occur chiefly in myeloid leukaemias and are not birefringent.

529. A Gaucher cell in the bone marrow showing typical coarse, onion skin, lipid inclusion material.

530. Two further Gaucher cells in the bone marrow – one typical cell and one heavily vacuolated.

529

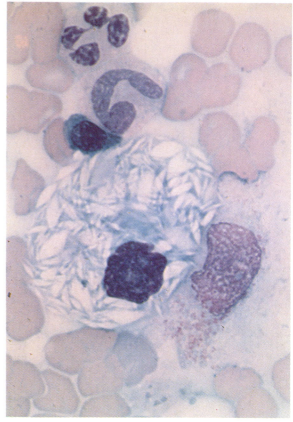

528

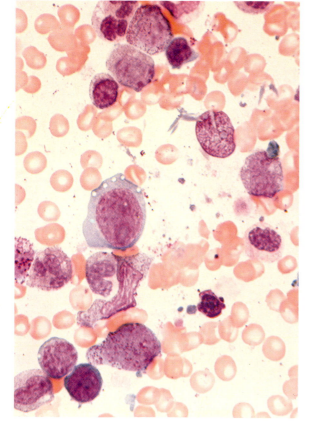

530

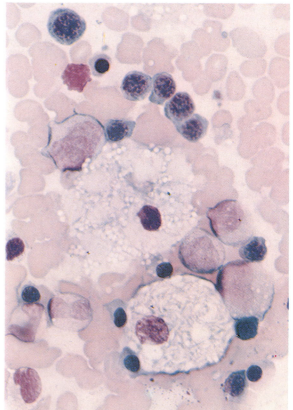

531. Less typical Gaucher cells from the same specimen. The cytoplasm is more granular and less fibrillary.

532 and 533. Gaucher cells under polarised light: the birefringence is shown in the same cells with the polariser turned to a 90° angle.

534. RE cells laden with foamy deposits of abnormal lipid in the bone marrow from a patient with Niemann-Pick disease.

532

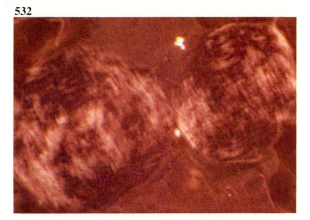

533

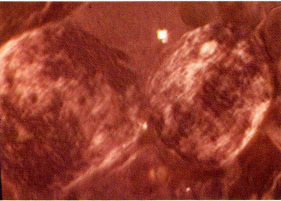

531

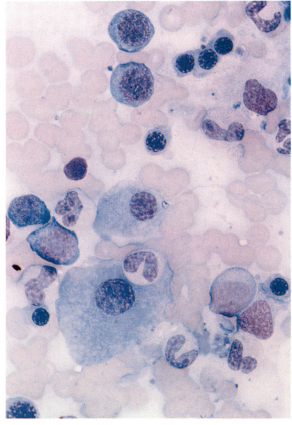

534

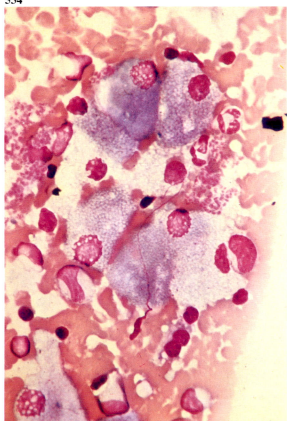

535. Abnormal cells, probably reticulo-endothelial macrophages, with variable reticulated cytoplasmic content of abnormal storage material, from the marrow of a patient with a 'histiocytosis' of the eosinophil granuloma-Hand-Schuller-Christian disease group.

536. Marrow cells from an unusual case of histiocytic leukaemia, showing general cytological similarities to the cells depicted in **535**. Phagocytic inclusions are conspicuous.

537 and 538. Further fields from the marrow of the same case of histiocytic leukaemia, showing cells of primitive cytology, with leptochromatic nuclei and variable basophilic cytoplasm, but with scattered vacuoles, occasional multiple nuclei and frequent giant nucleoli.

536

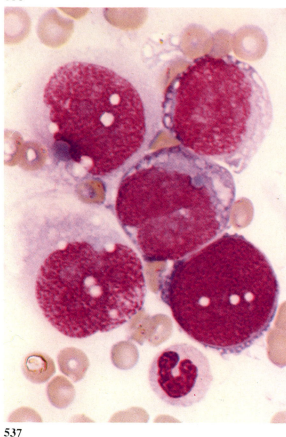

535

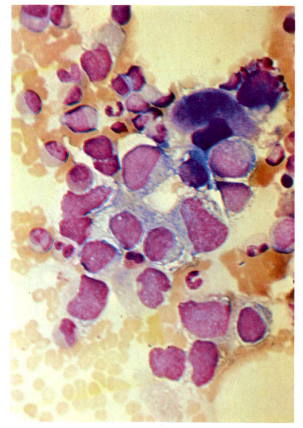

537

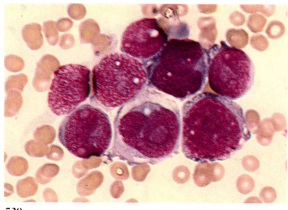

538

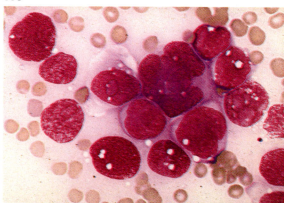

539. Sudan black reaction shows these primitive cells to be negative.

540. The PAS reaction shows remarkably strong positivity in virtually every cell.

541. Acid phosphatase is moderately strong in the blast cells, scattered over the nucleus and with local concentration in one cell at the site of an ingested red cell.

542. Dual esterase reaction: the blast cells show BE positivity with some fine CE granules. Two segmented cells show normal CE positivity and strong BE reaction is shown in a mature RE cell.

543 and 544. Free iron stain showing positive reaction in some of the primitive histiocytic cells.

541

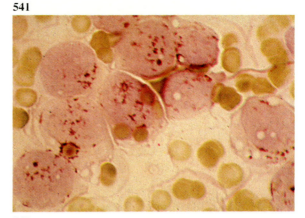

542

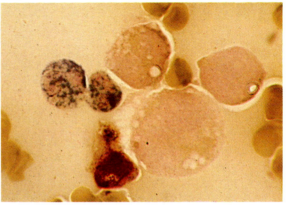

539

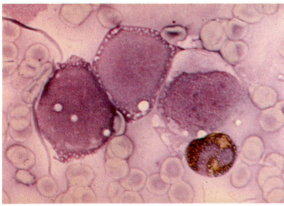

540

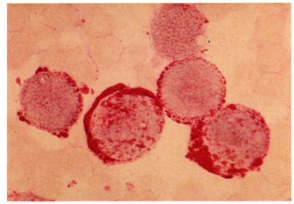

543

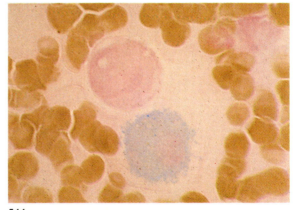

544

545. Malignant histiocytosis (histiocytic medullary reticulosis) showing erythrophagocytosis by bone marrow RE cell (histiocyte).

546. Free iron stain on similar malignant phagocytosing RE cells.

547. A further example of increased phagocytic activity in malignant histiocytosis. The whole clump of erythroblasts, showing faint diffuse PAS positivity, is within the cytoplasm of a spread RE cell with nucleus to the left of the clump.

546

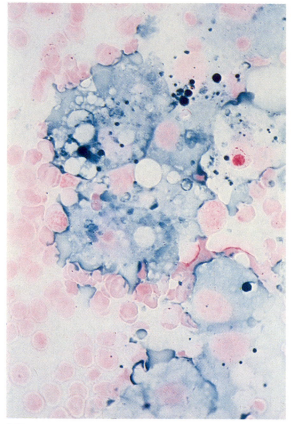

545

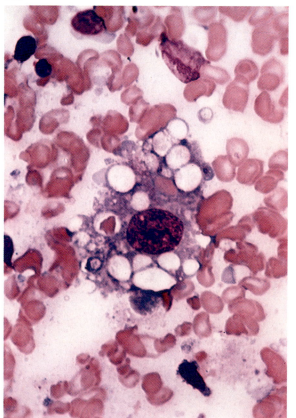

547

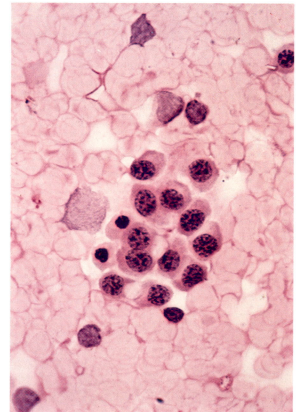

548. An osteoclast from a smear of normal bone marrow. The multiple, discrete, round or oval nuclei with their single conspicuous nucleoli, and the pale blue, lightly granular cytoplasm are characteristic and prevent confusion with megakaryocytes.

549. A group of osteoblasts, from a smear of normal bone marrow. They have a superficial resemblance to plasma cells, but are larger, do not show the heavy, 'clock-face', nuclear chromatin disposition seen in mature plasma cells, yet have a low nuclear-cytoplasmic ratio, with abundant, rather pale and filmy cytoplasm.

550. A further group of osteoblasts, whose large size is emphasized by the neighbouring segmented neutrophil.

549

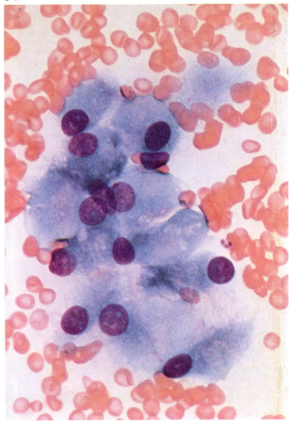

548

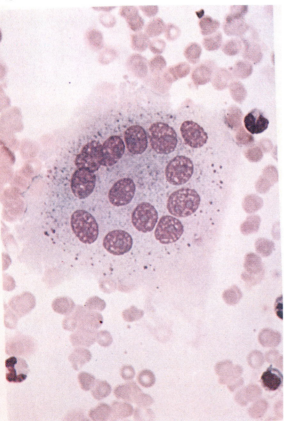

550

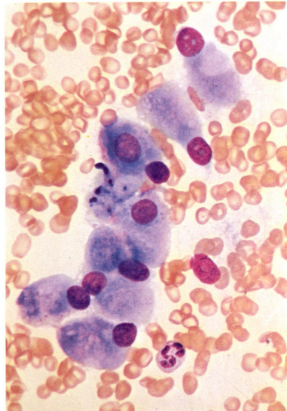

551. Scattered PAS positivity in osteoblasts.

552 and 553. Strong positivity for alkaline phosphatase in osteoblasts. This provides another differential point from plasma cells, which are negative for this enzyme.

554 and 555. Weak to moderate positivity for acid phosphatase in osteoblasts. This reaction is much stronger in plasma cells.

552

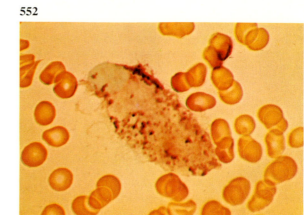

553

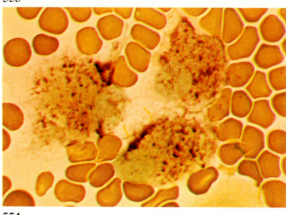

554

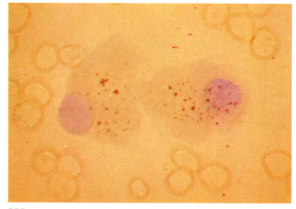

555

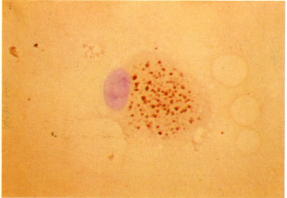

551

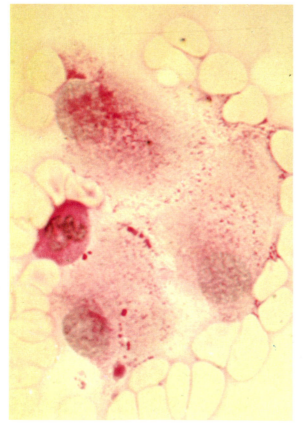

556 and 557. Examples of tissue mast cells in bone marrow smears. These cells have been illustrated earlier in macroglobulinaemia, where they are usually to be seen, but they also occur in small numbers in normal marrow and may show an increase in aplastic and hypoplastic states.

558. PAS reaction shows strong positivity in a mast cell.

559. Dual esterase reaction showing strong chloroacetate esterase positivity in a mast cell.

557

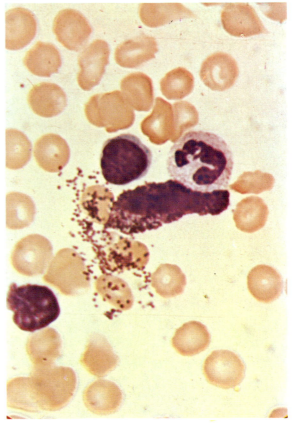

556

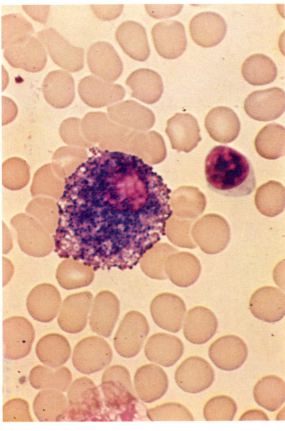

558

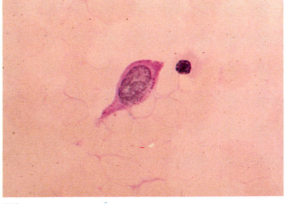

559

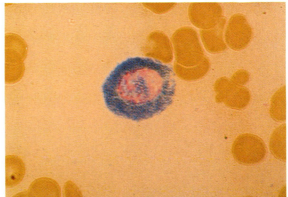

560–564. Examples of vascular endothelial cell clumps in the peripheral blood. The very regular nuclear structure and size and the tendency for these cells to occur in sheets or streaks along the direction of spreading of the film assist in recognition. These cells are foreign to blood, and are lifted from the intima of the vein during insertion or withdrawal of the needle used for collecting the blood sample. The nuclei in **563** both show conspicuous Barr bodies, indicating the presence of the inactive X-chromosomal material of normal female cells. Figure **564** shows alkaline phosphatase positivity in a strand of vascular endothelium, crossing a field of negative immature marrow cells.

561

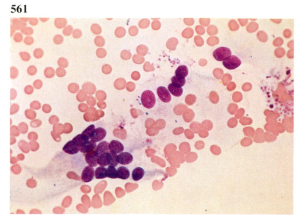

562

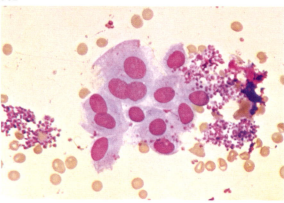

560

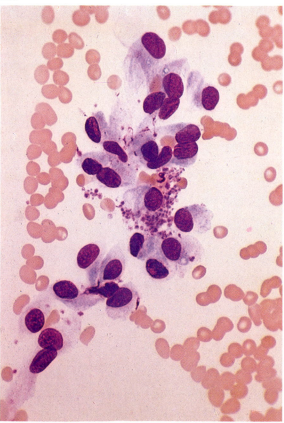

563

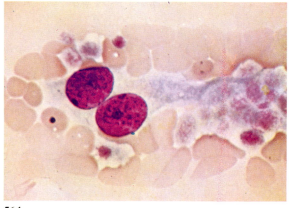

564

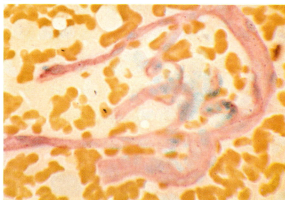

565. Droplet contamination. An artefact produced by coughing over an unfixed blood smear before staining.

566. A higher magnification reveals a cell from the buccal mucosa, with heavy bacterial contamination.

567. One intact promyelocyte and two flattened and partially disrupted promyelocytes, showing exaggerated nucleoli and open nuclear network. This appearance has been known as a 'Ferrata' stage of degeneration.

568. Two buccal mucosal cells with contained bacteria in a bone marrow smear.

567

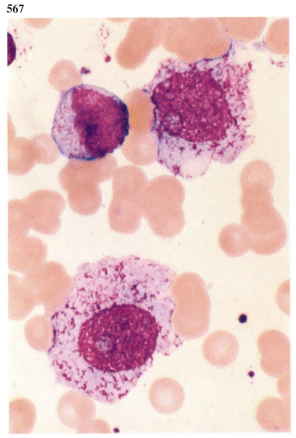

565

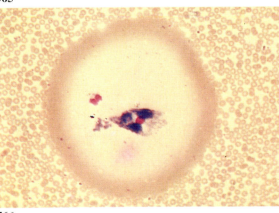

566

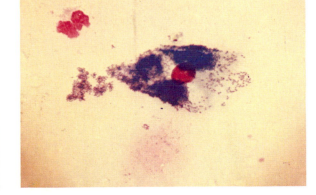

568

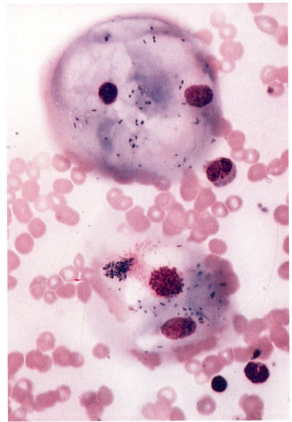

569. A fat-laden RE cell or lipophage from normal bone marrow.

570. A sebaceous skin cell contaminating a marrow smear.

571 and 572. Stromal fat cells in bone marrow smears.

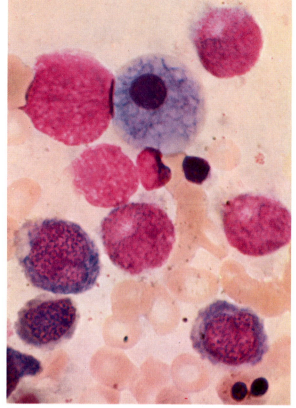

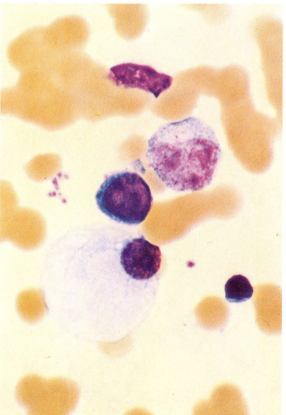

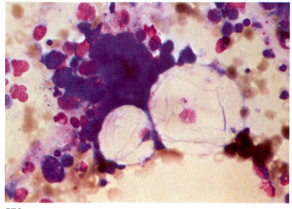

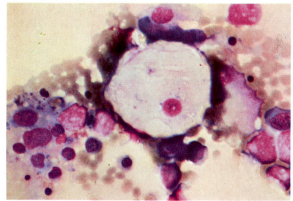

573 and 574. Examples from two different subjects, of neuroblastoma cell nests in the bone marrow. The cells have a resemblance individually to lymphoblasts of acute leukaemia, but frequently show a whorled arrangement in small clumps, as illustrated here.

575. PAS reaction on a clump of neuroblastoma cells, showing negative reaction, contrasting with the usual coarse positivity seen in a proportion of lymphoblasts in ALL.

574

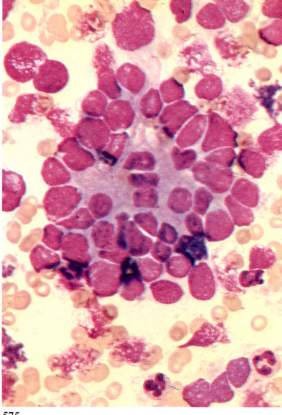

573

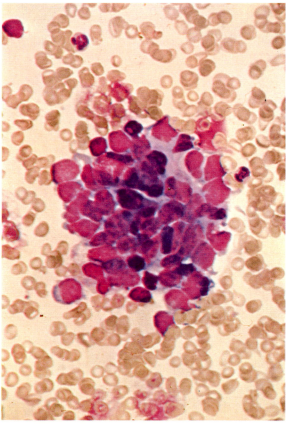

575

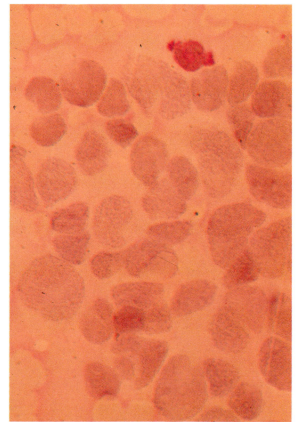

576–581. *Chemodectoma (paraganglioma) cells in marrow.*

576 and 577. Leishman stain: the malignant cells show conspicuous nucleoli, sometimes multiple nuclei and generally a large amount of clear cytoplasm.

578. Sudan black with a normally reacting polymorph and negative chemodectoma cells.

579. PAS reaction in the same preparation showing negative or faint diffusely positive reaction in tumour cells.

580. Alkaline phosphatase reaction is negative in the tumour cells (in contrast to RE cells). A positive polymorph is present.

581. Acid phosphatase reaction shows moderately coarse scattered granular positivity in tumour cells.

578

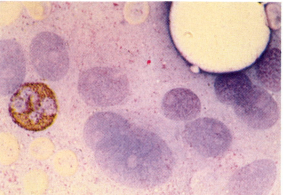

579

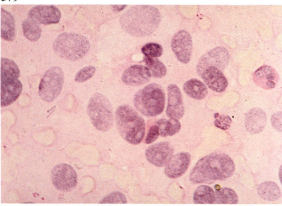

576

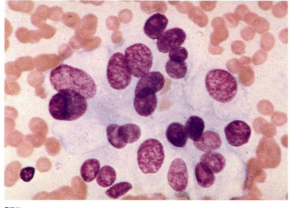

580

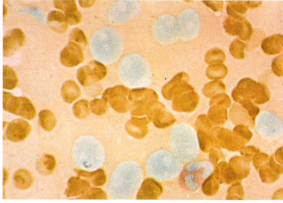

577

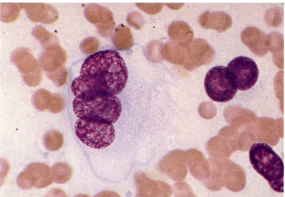

581

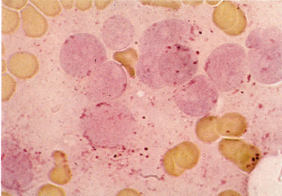

582. Clumped cells, darkly staining, and unlike any cells normally found in the marrow, allow a diagnosis of tumour cell metastasis to be made. This marrow smear was from a patient with disseminated carcinoma of the stomach.

583 and 584. Cytological detail of cell clumps from the same patient with gastric carcinoma. The tendency to syncytium formation is evident.

585. A partially syncytial cell clump of secondary deposit from carcinoma of the bronchus in a bone marrow smear.

583

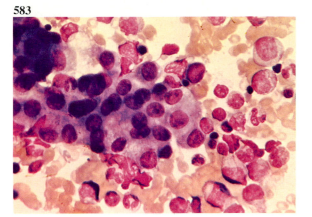

584

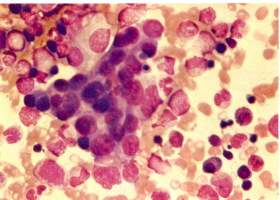

582

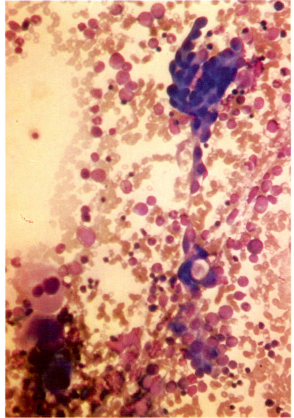

585

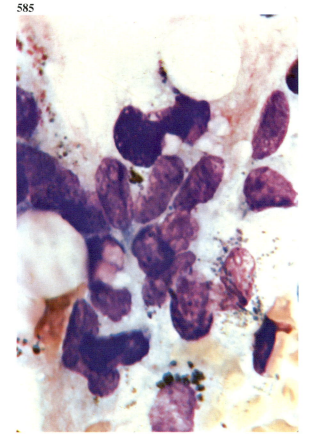

586. A further example of bronchial carcinoma cells in marrow, showing almost nothing but tumour cells (and clumped erythrocytes). A large multinucleated syncytium is present, perhaps of tumour cells, but possibly an osteoclast.

587. Bronchial carcinoma cells in marrow from a third patient, here showing a tendency to rosette formation.

588–590. Examples of malignant cells in bone marrow aspirates from a patient with disseminated carcinoma of the breast.

588

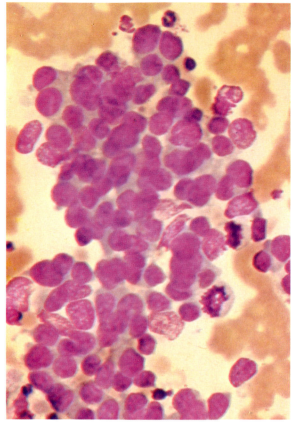

586

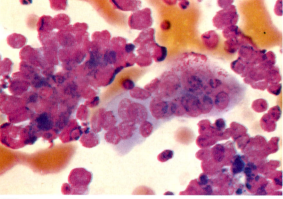

587

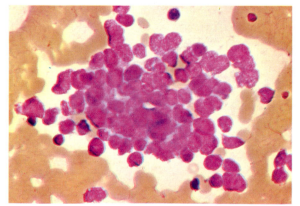

589

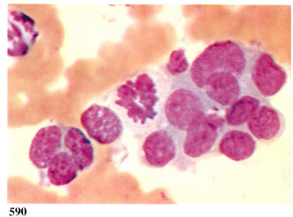

590

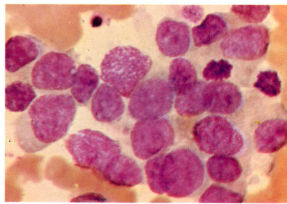

591–594. Giant cells (Reed-Sternberg cells) from the bone marrow of a patient with advanced Hodgkin's disease. They share the conspicuous, large, blue nucleoli which are very characteristic of Hodgkin's disease giant cells in smears or imprints of lymph nodes. Figures **593** and **594** are duplicate Romanowsky and consecutive PAS stains showing the generally weak positivity which most R–S cells manifest.

595. Involvement of both marrow and peripheral blood in the terminal stage of Hodgkin's disease may produce a leukaemia-like picture. This does not occur commonly, but most examples reported appear to have shown monocytoid primitive cells. Here is illustrated the peripheral blood picture in a terminal leukaemic transformation of Hodgkin's disease, where the cells probably belong to the same lineage as Reed-Sternberg cells.

593

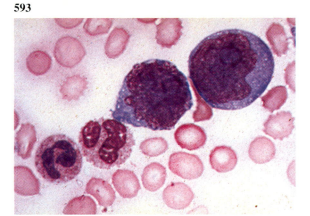

594

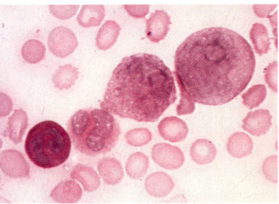

591

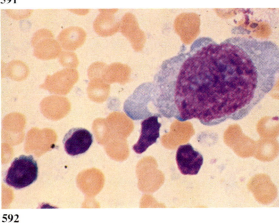

592

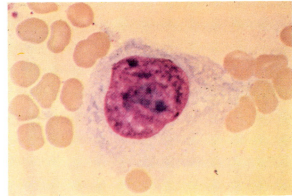

595

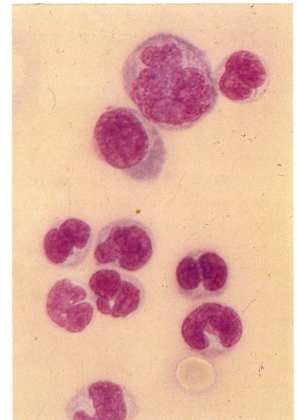

596. Blood smear in infection with Bartonella bacilliformis. Many erythrocytes contain the organisms, with rod-shaped or sometimes cocco-bacillary form. The disease produced by this organism in man, Bartonellosis, Carrion's disease or Oroya fever, is associated with macrocytosis and haemolytic anaemia, and this smear shows some polychromasia.

597. Yeast-like bodies of blastomycosis in a spicule of bone marrow, possibly undergoing phagocytosis by reticulo-endothelial cells.

598. A reticulo-endothelial cell containing many Leishman-Donovan bodies, from the marrow of a patient with kala-azar. Various other marrow cells, a mega-karyocyte, several myelocytes and a late normoblast, are also present.

597

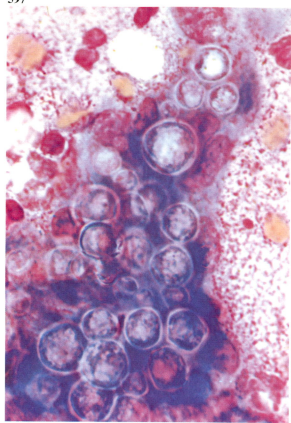

596

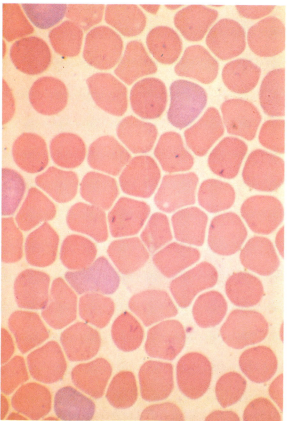

598

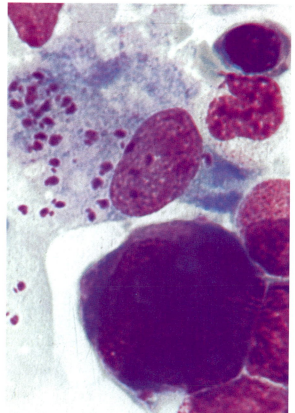

599. Another example of Leishmania, with organisms within a monocyte and others liberated among the surrounding cells. Below the monocyte is a clump of platelets for comparison with the more sharply and positively staining Leishmania bodies.

600. Further Leishmania parasites in a monocyte/macrophage.

601. Toxoplasma within the cytoplasm of a macrophage. Two free organisms are present towards the bottom right, and other single free organisms elsewhere. Their larger size, spindle shape and banded nucleus distinguishes them clearly from Leishmania and from platelets.

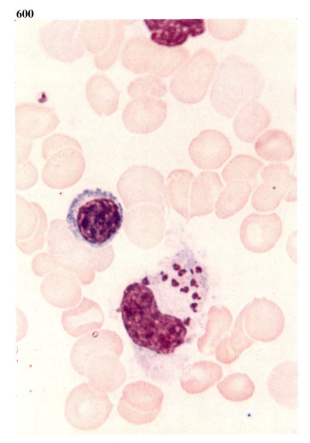

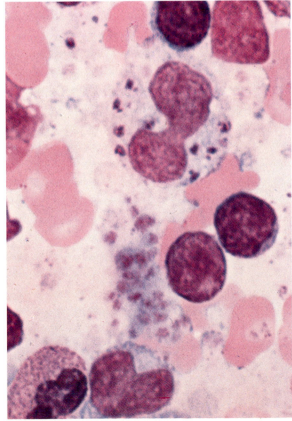

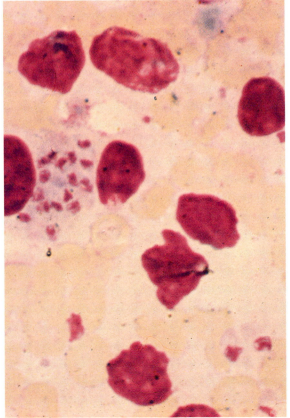

602–604. *Plasmodium falciparum infection.*

602. Ring form trophozoites in erythrocytes. The presence of several rings in a single erythrocyte is strongly suggestive of P. falciparum.

603. Further ring forms and a mature trophozoite approaching division.

604. Crescentic gametocytes, diagnostic of P. falciparum.

603

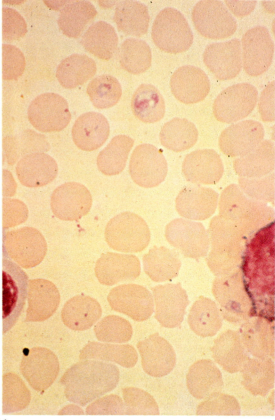

602

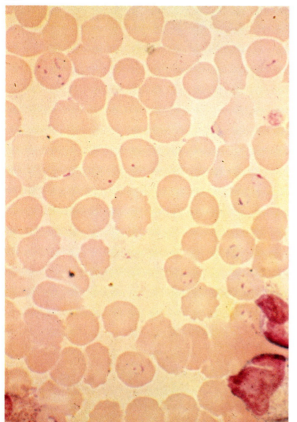

604

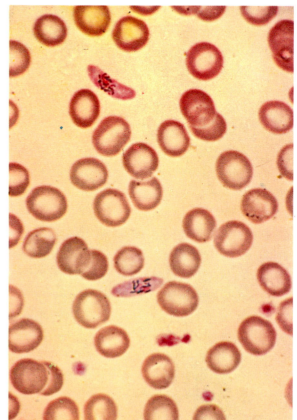

605–608. *Plasmodium malariae infection.*

605. Ring form trophozoites in erythrocytes.

606. An example of the coarse ring forms found in P. malariae infections (and also in P. vivax and P. ovale, but not in P. falciparum).

607. Maturing trophozoite.

608. Schizonts containing between 6 and 12 round merozoites, together with clumps of coarse dark-brown pigment.

606

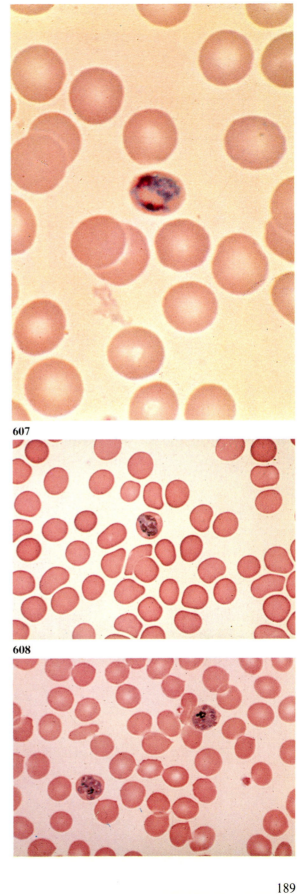

605

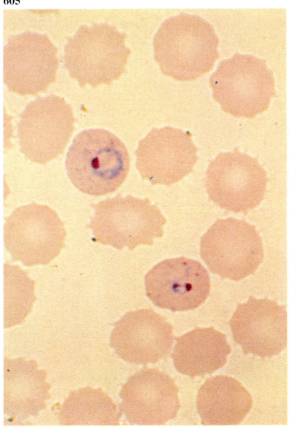

607

608

189

609. Ring forms in P. vivax infection.

610. Ring form and more mature trophozoites, from P. ovale infection. Schuffner's dots in the erythrocytes are conspicuous, as they may also be in P. vivax infections.

611. Schizont in P. vivax infection, containing some 20 merozoites and a quantity of dense brown pigment.

612. P. vivax infection. Ring form trophozoite, and liberation of merozoites from a fully developed schizont.

610

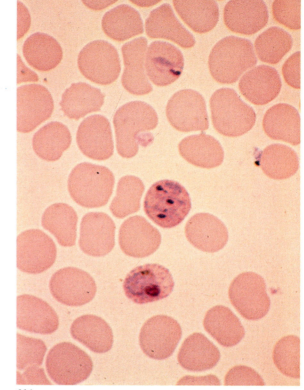

609

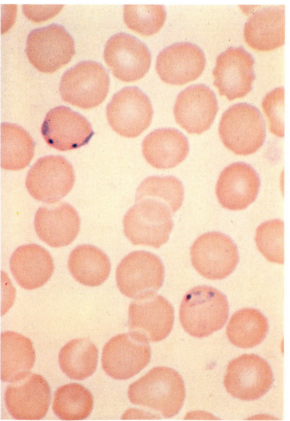

611

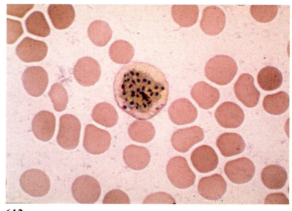

612

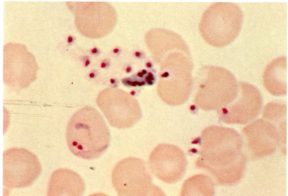

613. Trypanosoma gambiense, the most important species causing African trypanosomiasis. The kinetoplast is much smaller than the central nucleus, the undulating membrane is broad and the contortions of shape variable.

614. Trypanosoma brucei, present in some areas of Africa, has generally similar morphology.

615. Trypanosoma cruzi, the cause of Chagas' disease or South American trypanosomiasis, has a much larger kinetoplast, a less conspicuous undulating membrane and a more fixed horse-shoe configuration.

614

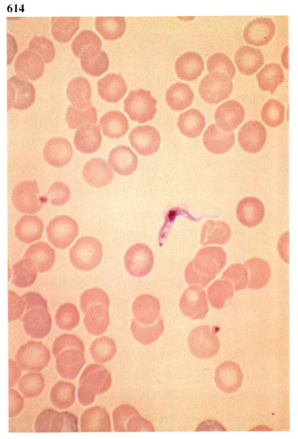

613

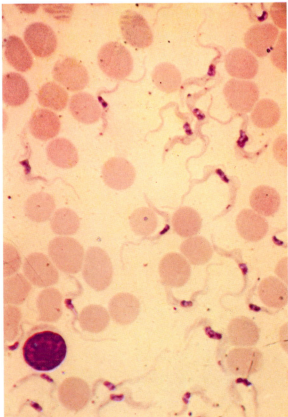

615

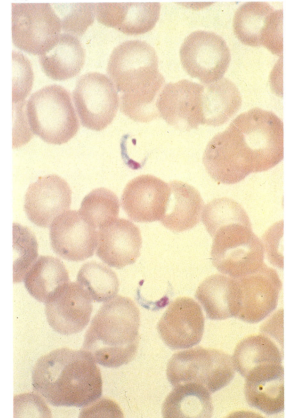

Part 5
Imprints of lymph nodes and spleen; cells from pleural, ascitic and cerebrospinal fluids

Cytological study of imprint or needle aspirate samples of lymph nodes and spleen complement histological study of biopsy specimens and are especially valuable in the diagnosis of infective and reactive lymphadenopathy, lymphomas and some other metastatic tumours, and, in the case of splenic material, also in lipidoses, malignant histiocytosis and some primarily haematological disorders with extramedullary haemopoiesis. The nomenclature of lymph node cells is confusing and several systems exist. In the illustrations here we give certain of the more widely used synonyms initially and then use chiefly the simple Kiel classification, expecting readers to translate as necessary. The cytological differences between lymphomas, for example, are generally clear in the slides depicted, even if semantic dispute remains.

Phagocytic reticulo-endothelial cells (RE cells) in bone marrow and blood have been illustrated previously; similar cells are conspicuous in lymph nodes where their development from monocytes through an intermediate monocyte/macrophage or epithelioid cell may be envisaged. Cells of the same family – perhaps the same cells in different topographical sites or at different functional stages of activity – have been separated by histologists and electron microscopists into a series of types. Apart from the actively phagocytic RE cell (histiocyte, macrophage, starry sky cell, histiocytic reticulum cell) and the epithelioid cells, fibroblastic, dendritic and interdigitating RE or reticulum cells have been described. These latter are not often clearly distinguishable in imprints, perhaps because they tend to remain in the supporting structure of the node and do not easily come free in touch or smear preparations. The RE cells that do appear in imprints are shown in Romanowsky preparations and also with various cytochemical stains which often render them especially conspicuous. Such stains may help in identifying the variants listed above – fibroblastic RE cells are described as strongly positive for alkaline phosphatase but only weakly so for esterases and acid phosphatase; dendritic cells, mostly from germinal centres, are negative for phosphatases and weak in esterase; interdigitating cells, chiefly from extrafollicular T-cell zones, are negative for alkaline phosphatase and react weakly for acid phosphatase and esterase. Our own impression is that these criteria are no more than rough guidelines; certainly the cytochemical positivity of clearly phagocytic RE cells varies widely and is probably much influenced by the nature of the ingested cellular material. The degree of variability of reaction in the cell group as seen in imprints is illustrated in this section.

We have not attempted to illustrate a wide range of secondary tumour metastases in lymph nodes, but only a few examples where clinical confusion with lymphoma existed at the time of biopsy.

The occurrence of leukaemic or lymphomatous infiltration in pleura, peritoneum or meninges, may lead to the presence of neoplastic cells in pleural, ascitic or cerebrospinal fluids. The neoplastic cells, which may be seen in centrifuged deposits from these fluids, look much as they do in buffy coat preparations from blood or in marrow, as illustrated already in earlier sections of this atlas. They are accordingly not extensively shown again here, but some illustrations are given of a single example of immunoblastic lymphoma and several of non-malignant cells including lining cells and reactive immunoblasts which must be differentiated from neoplastic ones in these circumstances.

616. Leishman stain: lymph node imprint: showing three large immunoblasts or activated, transformed lymphocytes, with ample basophilic cytoplasm, and generally primitive nuclei with nucleoli more centrally than peripherally situated; a somewhat smaller centroblast or large follicular centre cell, top left, with higher nuclear-cytoplasmic ratio but also with cytoplasmic basophilia and three nucleoli more towards the periphery of the nucleus; mature lymphocytes appear above and below the centroblast near the lowest immunoblast; there is a single plasma cell near the central immunoblast and a monocyte/macrophage/RE cell with a small inclusion at the left of the field. The remaining cells are centrocytes or small follicular centre cells.

617. Leishman stain: lymph node imprint: showing seven cells of the immunoblast or activated lymphocyte type to illustrate the range of size and cytology manifest in this group; although nuclear chromatin shows considerable variation in density all cells have distinguishable nucleoli and ample basophilic cytoplasm. The remaining cells in this field are mostly mature lymphocytes.

618. Leishman stain: lymph node imprint: a collection of cells from the follicular centre area of a lymph node imprint in reactive hyperplasia secondary to infection. Against a background of small mature lymphocytes are occasional small follicular centre cells (centrocytes) and about a dozen large FCCs or centroblasts with nucleoli and some cytoplasmic basophilia. In the upper right corner are several immunoblasts, and to their left a histiocytic RE cell and a plasma cell.

617

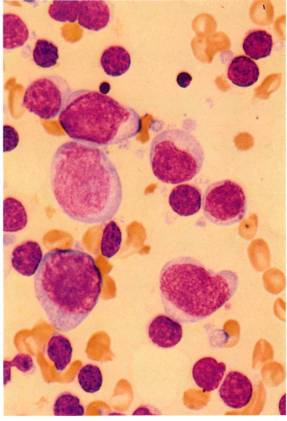

616

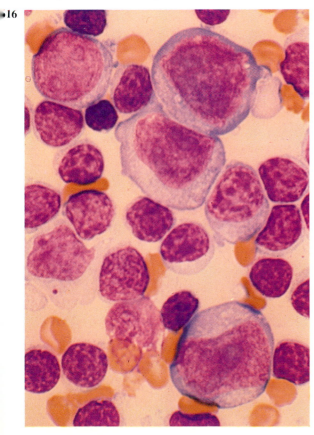

618

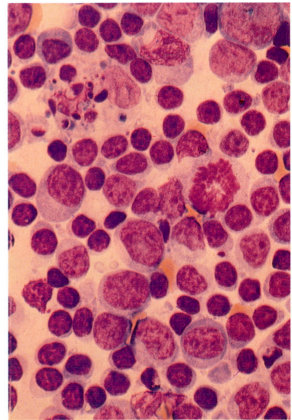

193

619. Leishman stain: lymph node imprint: field to illustrate seven large cells involved in a sequence of probable developmental changes along the monocyte-epithelioid cell-macrophage (RE cell) line. The remaining cells in this field are chiefly small lymphocytes.

620. Leishman stain: lymph node imprint: an example of an epithelioid cell cluster in a lymph node showing a reactive hyperplasia secondary to toxoplasma infection. Around the cluster are chiefly lymphocytes but with a histiocytic RE cell (starry sky cell) at the bottom and six large mononuclear cells, one in mitosis, at the centre and lower right of the field. These cells are probably immunoblasts.

621. Leishman stain: lymph node imprint: RE cells, perhaps of the interdigitating variety, with little evidence of phagocytosis, but with neighbouring lymphocytes, occupy much of the lower half of this field. Two are binucleated, the left with an adjacent probable tissue mast cell and the right with a more basophilic cytoplasm than usual. The upper half shows a mixture of small lymphocytes with dense nuclei and follicular centre cells or centrocytes with more open nuclei, showing transition towards centroblasts and with a single immunoblast towards the top right corner.

620

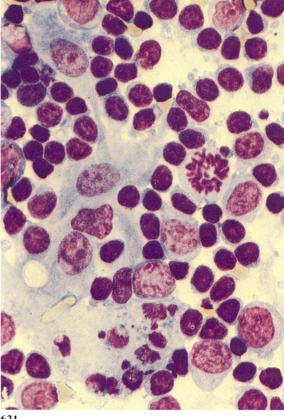

619

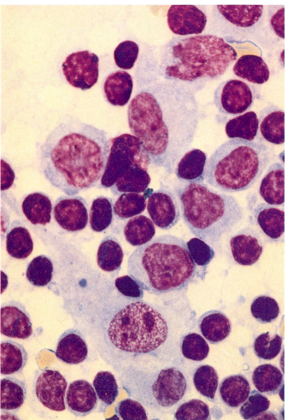

621

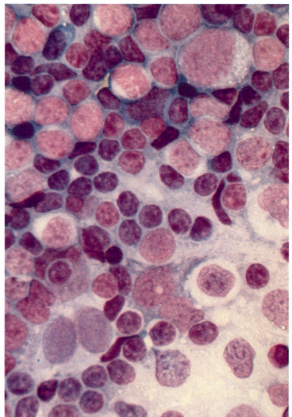

622. Sudan black stain: lymph node imprint: Sudan black positivity is shown in a histiocytic RE cell and in three cells of the monocyte-macrophage line. All lymphoid elements are negative.

623. Sudan black stain: lymph node imprint: positivity is shown in a neutrophil polymorph and metachromatic staining in a tissue mast cell. The surrounding centrocytes, centroblasts and immunoblasts are all SB negative.

624. PAS reaction: lymph node imprint: this preparation from the same node as in **621** shows fine granular positivity in a few mononuclear cells only. One on the left is probably an RE cell of the interdigitating type with granular PAS positivity spreading out in the cytoplasmic prolongations between neighbouring cells. Two other positive cells towards the top of the field are respectively a centroblast and a centrocyte. A very occasional lymphocyte also shows detectable granular positivity.

623

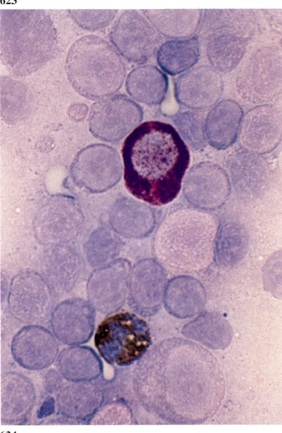

622

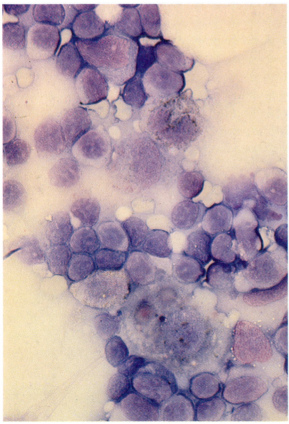

624

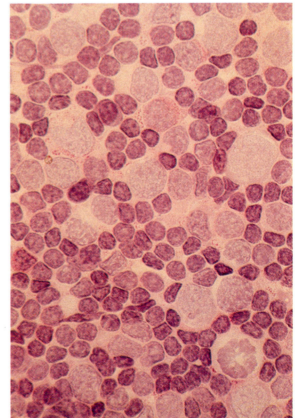

625. PAS reaction: lymph node imprint: this preparation shows normally positive reactions in neutrophil polymorphs but generally negative reactions in lymphocytes, centrocytes and centroblasts, and no more than very weak diffuse or finely granular positivity in five RE cells of differing morphology. One of these contains an ingested polymorph and a lymphocyte and another has four nuclei.

626. Alkaline phosphatase: lymph node imprint: this preparation from a reactive node shows lymphocytes, centrocytes, centroblasts and immunoblasts all to be negative for this enzyme, while a single neutrophil polymorph shows + + positivity and a large phagocytic RE cell is only very weakly positive despite its obvious past ingestion of various other cell types, probably including a neutrophil.

627. Alkaline phosphatase: lymph node imprint: strong positivity is manifest in a phagocytic RE cell with ingested cellular remnants; positively reacting fibrils spread out over a neighbouring lymphocyte to give it a spurious appearance of positivity.

626

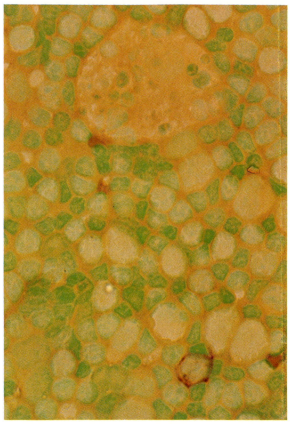

625

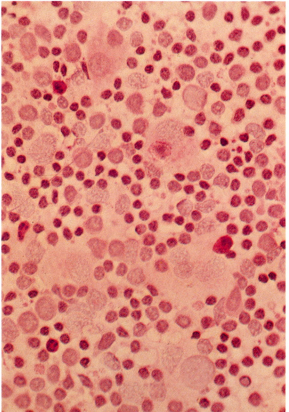

627

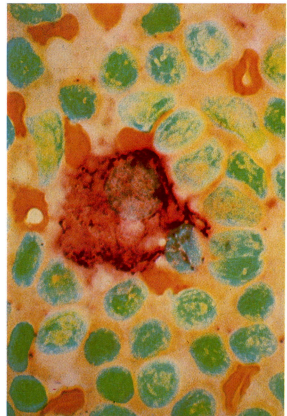

628. Acid phosphatase: lymph node imprint: very strong positivity is seen in a phagocytic RE cell containing much ingested material, while at the opposite corner of the field a smaller monocyte-macrophage shows less intense but still striking positivity. Most of the remaining cells are of the lymphocyte, centrocyte, centroblast family and show no more than occasional weak granular positivity, but between the cells spreads a network of positively reacting fibrils probably derived from the larger histiocytic RE cell or perhaps from an inconspicuous dendritic RE cell.

629. Acid phosphatase: lymph node imprint: several strongly positive RE cells of variable size and shape are seen against a background of largely negative lymphocytes. An occasional probable T cell with coarse cytoplasmic granules is present, and also several normally positive neutrophil polymorphs.

630. Acid phosphatase: lymph node imprint: a phagocytic or histiocytic RE cell with much ingested material (starry sky cell) shows relatively little acid phosphatase positivity in contrast to a more strongly positive monocyte or epithelioid cell. Lymphocytes here show mostly either negative or scattered granular reactions and there are two large mononuclear cells, with variable weak granular positivity, which are probably immunoblasts. At the lower right corner is a positively reacting plasma cell.

629

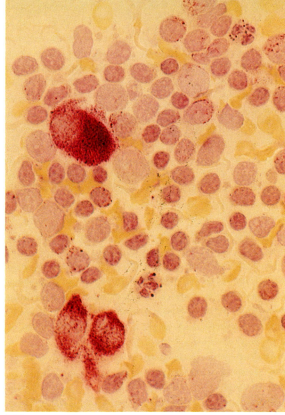

628

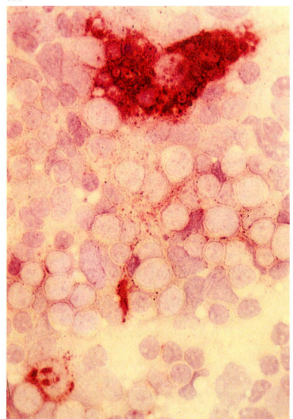

630

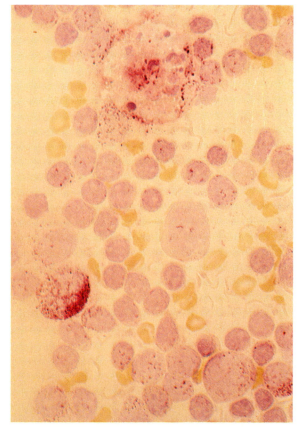

631. Dual esterase reaction: lymph node imprint: this preparation shows chiefly cortical area Tμ lymphocytes, with strong localised dots of BE positivity, surrounding larger cells of the monocyte-macrophage system only one of which shows moderately strong BE positivity, the remainder being almost negative with only a few CE-positive granules. A normally CE-positive polymorph is present.

632. Dual esterase reaction: lymph node imprint: BE positivity of moderate to strong degree in epithelioid cells in this case. One has ingested a CE-positive neutrophil. Most lymphocytes here do not show the localised Tμ type of BE positivity and are presumably chiefly B cells.

633 and 634. Dual esterase reaction: lymph node imprint: these fields from the same case of reactive hyperplasia show strong BE positivity in RE cells probably of the histiocytic or phagocytic variety, weaker reactions in occasional monocyte-macrophage cells and localised dot-like positivity in certain lymphocytes, presumably of the Tμ subset. Figure **634** shows also a probable dendritic RE cell with long intercellular processes but rather weak BE positivity.

632
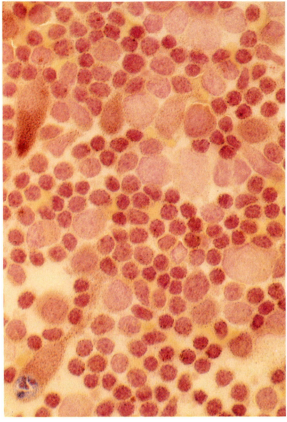

631
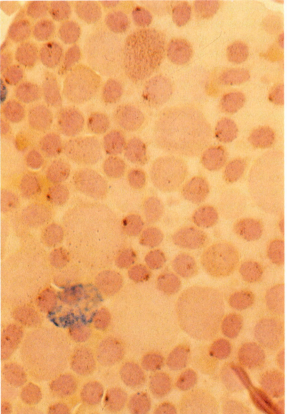

633
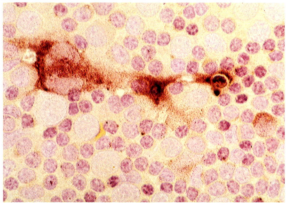

634
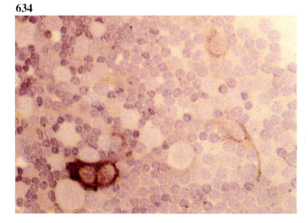

635–637. *Leishman stain: lymph node imprints.*

635. Group of epithelioid cells with some phagocytic activity surrounded by lymphocytes and centrocytes from the chronic lymphadenitis of sarcoidosis.

636. A foreign body giant cell or Langhans cell from the same condition.

637. Another example of a Langhans multinucleated giant cell from tuberculous lymphadenitis.

636

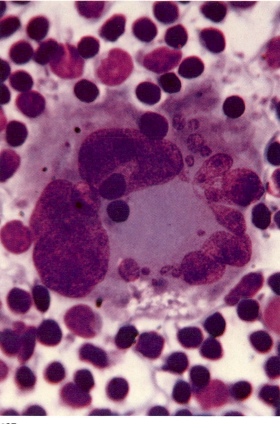

635

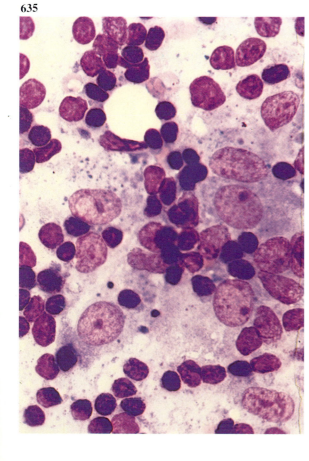

637

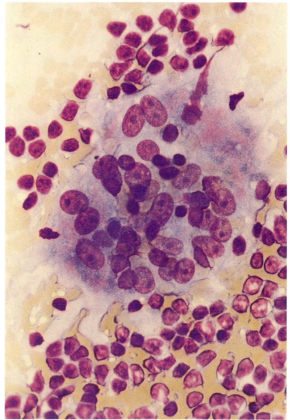

638–640. *Leishman stain: lymph node imprints: Hodgkin's disease – mixed cellularity.*

638 and 639. Low and higher power views showing the pleomorphic picture commonly seen in node imprints from the disease, especially in the mixed cellularity variant. Lymphocytes, centrocytes, occasional centroblasts and immunoblasts mingle with plasma cells, eosinophil and sometimes neutrophil polymorphs, and monocyte-macrophages. There is a single large Reed-Sternberg (R-S) cell with twisted or overlapping double nucleus and large dark violaceous nucleoli in each figure. The extent of eosinophilia in this example is unusual, but some eosinophils can commonly be found in most imprint preparations from Hodgkin's disease.

640. Another area of the same slide again showing lymphocytes, centrocytes, centroblasts and an immunoblast (top right) together with a plasma cell and many eosinophils. Centrally there is a macrophage or RE cell with on the right a mononuclear R-S cell or Hodgkin's cell with characteristic nuclear chromatin and nucleoli.

639

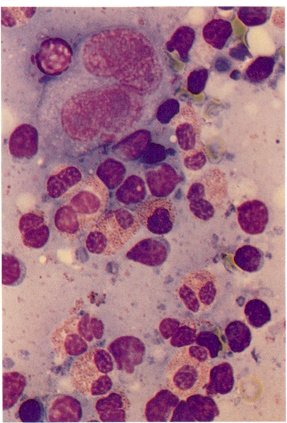

638

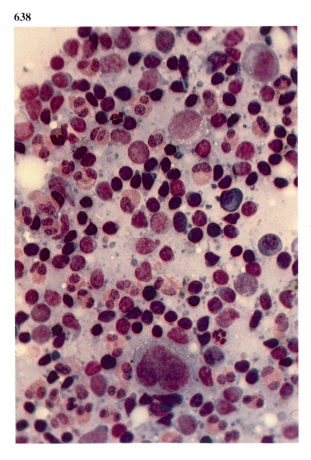

640

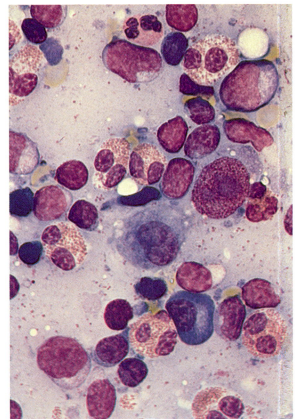

641–643. *Leishman stain: lymph node imprints: Hodgkin's disease – lymphocyte predominant.*

641. Low power view showing a more uniformly lymphocytic and centrocytic background, but with one binucleated R-S cell and several mononuclear Hodgkin's cells. Occasional monocyte-macrophage type cells can be seen.

642. A higher power view of the same slide in which all the cell types mentioned above can be more clearly identified.

643. A still higher magnification of an R-S cell from this specimen, to illustrate the typical reticular nuclear chromatin and the large dark-violet nucleoli. The surrounding cells are mostly lymphocytes.

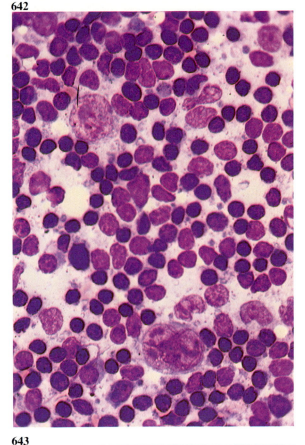

642

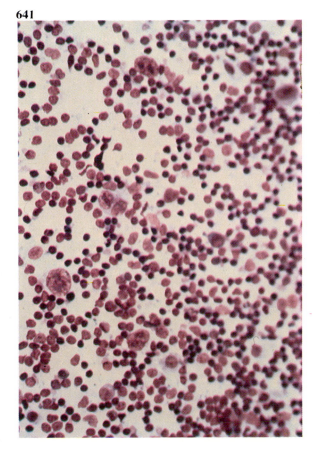

641

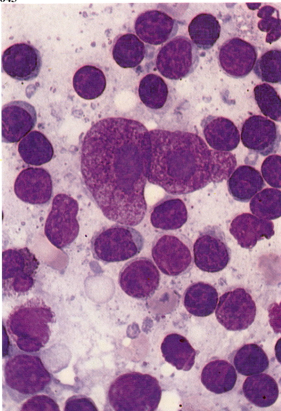

643

644–646. *Leishman stain: lymph node imprint: Hodgkin's disease – nodular sclerosing to mixed cellularity – morphological relationships of R-S cells.*

644. A field showing various binucleated R-S cells and mononuclear Hodgkin's giant cells. A centroblast at the lower right corner and a central immunoblast show morphological similarities to the smaller of the Hodgkin's cells and suggest a possible derivation of the R-S cells from activated B cells.

645. A second field from the same specimen allows similar parallels to be drawn: the R-S cell in the upper left corner relates to the probable mononuclear Hodgkin's cell in the lower right, and that in turn to the immunoblast above it.

646. Yet another area in the same slide shows a collection of histiocytic RE cells, some with ingested cell debris, which are clearly morphologically distinct from the R-S and Hodgkin's cells of the previous figures.

647–652. *Leishman stain: lymph node imprints: variants of giant cells in Hodgkin's disease.*

647–649. Typically binucleated R-S cells. There is also a mononuclear Hodgkin's cell in **649.**

646

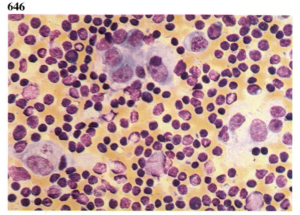

647

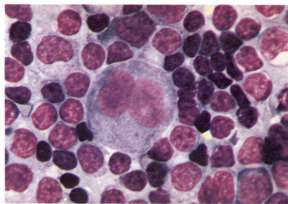

644

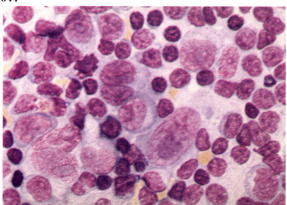

648

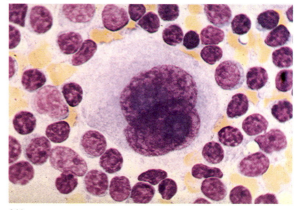

645

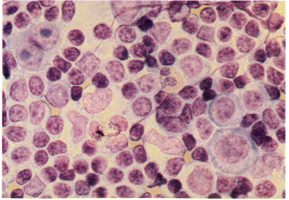

649

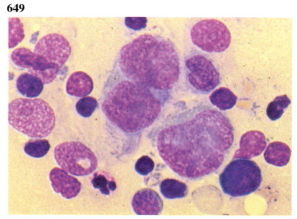

650–651. Mononuclear Hodgkin's cells, with characteristic nuclear and nucleolar patterns. In **651** a pair of these cells, one possibly with overlapping double nucleus, are separated by a pair of RE cells of clearly contrasting nuclear and cytoplasmic morphology.

652. Leishman stain: lymph node imprint. A mitotic figure in an R-S cell, showing the clearly intact nucleoli surrounded by separated chromosomes. To the left of this cell is a macrophage.

651

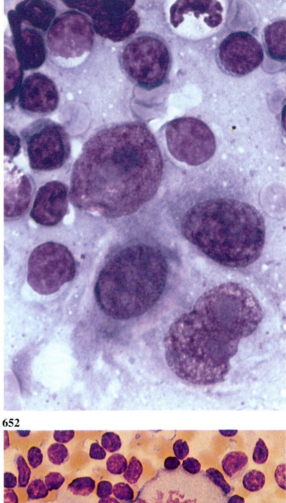

650

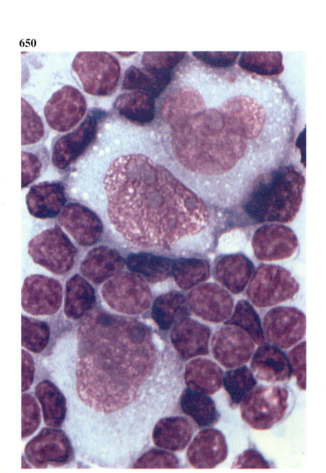

652

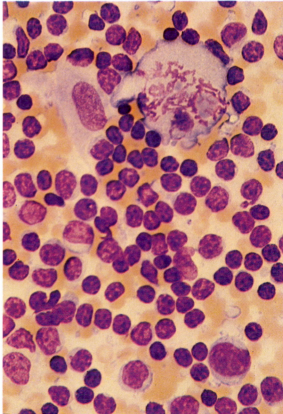

653. PAS reaction: lymph node imprint: Hodgkin's disease. A clearly binucleated R-S cell and five out of six mononuclear Hodgkin's cells, including one in mitosis, show globules or blocks of PAS positivity. Although the giant cells of Hodgkin's disease are most often PAS negative, the appearance of globules of positivity as in this figure, or even coarsely scattered granular positivity, is by no means rare.

654. Acid phosphatase stain: lymph node imprint: Hodgkin's disease. A binucleated R-S cell shows strong positivity.

655. Dual esterase reaction: lymph node imprint: Hodgkin's disease. Hodgkin's cells usually show little esterase positivity, as here, where weakly scattered BE and CE granules are detectable, in sharp contrast to the strong BE positivity in a neighbouring giant RE cell. An eosinophil, a CE-positive neutrophil and variably BE-positive lymphocytes complete the field.

654

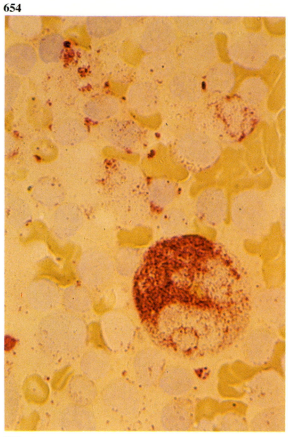

653

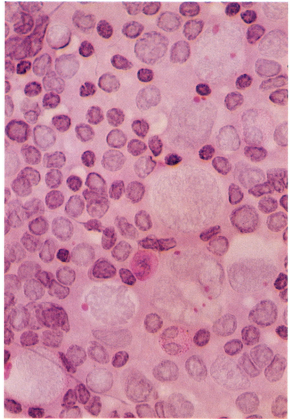

655

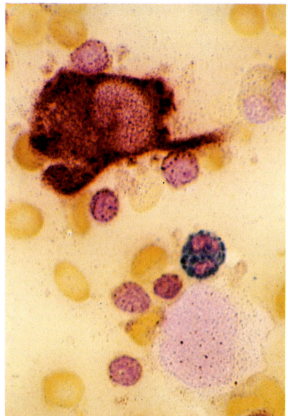

656 and 657. Leishman stain: lymph node imprint: chronic lymphadenitis. Multinucleated giant cells, with flat epithelial-type nuclei and relatively small or inconspicuous nucleoli; not to be mistaken for Hodgkin's disease giant cells.

658. Dual esterase reaction from the same specimen. One of the multinucleated giant cells shows strong BE positivity, unlike R-S cells which are usually negative. Localised dot-like positivity is seen also in some probable T cells and there is CE positivity in several neutrophils and mixed reaction in a plasma cell.

657

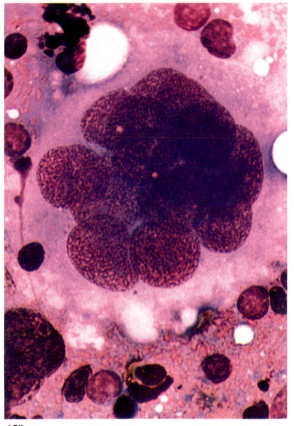

656

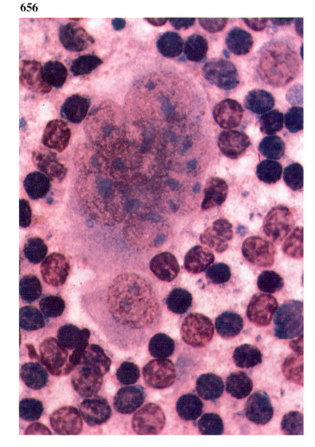

658

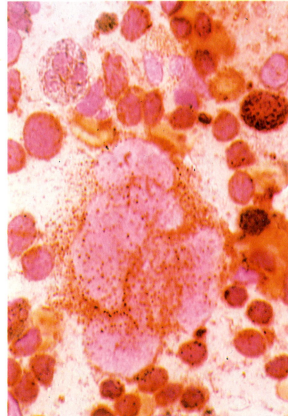

659–664. *Lymph node imprints: follicular lymphoma, centrocytic-centroblastic; mixed small and large follicular centre cell lymphoma.*

659. Leishman stain: lower power view shows a mixture of centrocytes and centroblasts with a scattering of mature lymphocytes.

660. A higher power view of the same preparation, showing large single or occasionally multiple nucleoli in the larger cells – centroblasts – which often show indentation or cleaving of the nucleus. The two varieties of small cell, densely pachychromatic lymphocytes and more lightly stained centrocytes can be clearly differentiated.

661. SB reaction in this node imprint shows all the lymphoma cells to be negative. The single positively reacting cell is a neutrophil polymorph.

660

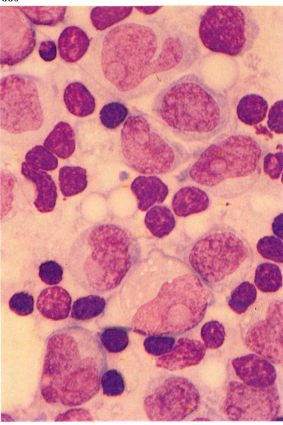

659

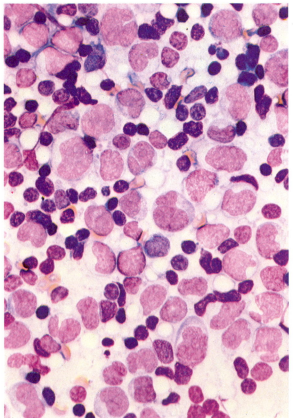

661

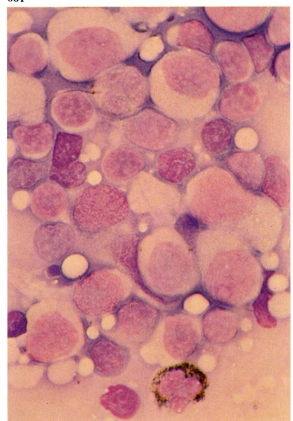

662. PAS reaction on the same node imprint shows a little positivity in cytoplasmic fragments and some tinge and fine granular reaction in lymphoma cells.

663. Acid phosphatase reaction on this specimen shows variable + to +++ positivity in lymphoma cells, scattered and not particularly localised in the paranuclear zone. There is a strongly positive RE cell present.

664. Dual esterase reaction here shows CE positivity in a neutrophil polymorph and in a tissue mast cell, with BE positivity of modest degree in an RE cell, possibly dendritic, with fine processes extending between neighbouring lymphoma cells. The lymphoma cells are essentially negative.

663
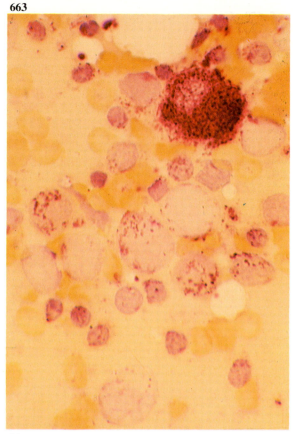

662
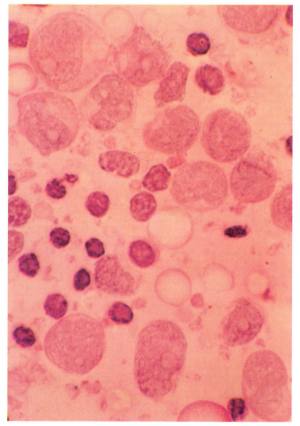

664
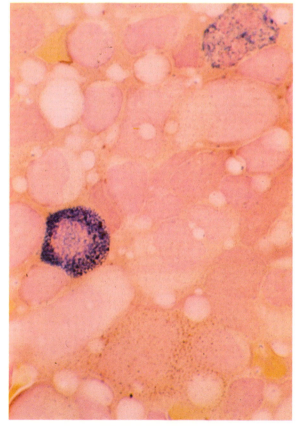

665–670. *Lymph node imprints: lymphoblastic lymphoma, B-cell type.*

665 and 666. Leishman stain: low and higher power views of the neoplastic lymphoblast tumour cells; they show rounded nuclei with moderately fine chromatin and nucleoli varying in number between one and three or four, mostly not at the nuclear membrane. There is moderately deep basophilia in the narrow cytoplasmic rim.

667. Sudan black: there is a single positive neutrophil polymorph but the lymphoblasts and occasional lymphocytes are all quite negative.

666

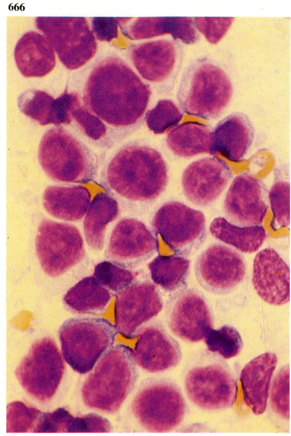

665

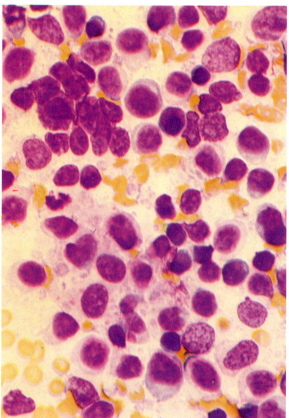

667

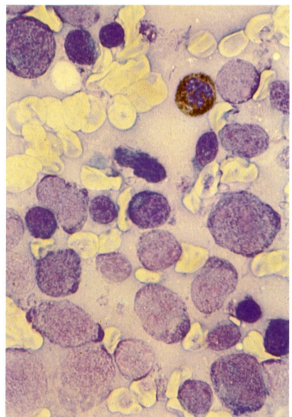

668. PAS reaction: the blast cells are virtually entirely negative; there is some granular positivity in one lymphocyte and the usual strong reaction in the single neutrophil present.

669. Acid phosphatase: the blast cells are again negative while a phagocytic RE cell is strongly positive. A few positive granules are seen in a single plasma cell above the RE cell.

670. Dual esterase: the lymphoblasts show fine scattered BE positivity. A neutrophil polymorph is CE positive.

669

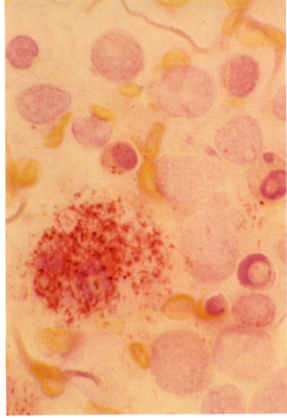

668

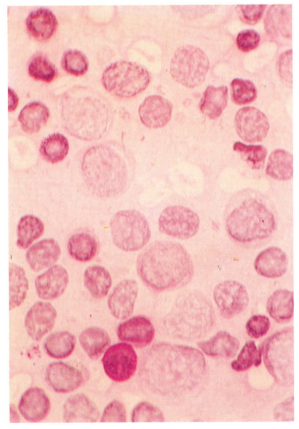

670

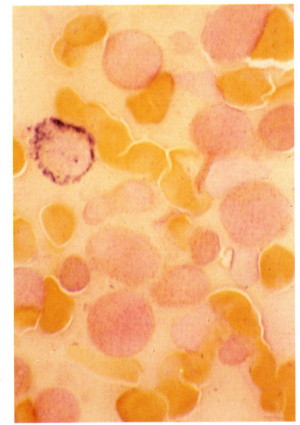

671–673. *Lymph node imprints: immunoblastic lymphoma, B-cell type, with some plasmacytoid differentiation.*

671 and 672. Leishman stain: low and high power views showing large immunoblastic cells with moderate amounts of basophilic cytoplasm and leptochromatic nuclei containing large but poorly defined central nucleoli. Most immunoblasts contain globular inclusions, staining a greyish-green colour, like Russell bodies, presumably representing secreted immunoglobulin. This feature is only occasionally seen in cases of immunoblastic lymphoma, but when present indicates the secretory B-cell nature of the tumour. Among the surrounding small cells, chiefly lymphocytes and centrocytes, are some with plasmocytic morphology occasionally containing similar Russell body inclusion material. Scattered cytoplasmic fragments are conspicuous especially in the high power view, where multiple small cytoplasmic vacuoles are visible in the immunoblasts.

673. PAS reaction shows the immunoblasts to be essentially negative, as are the globular inclusions in this instance, although more often such inclusions show weak PAS positivity. A single positive neutrophil can be seen.

672

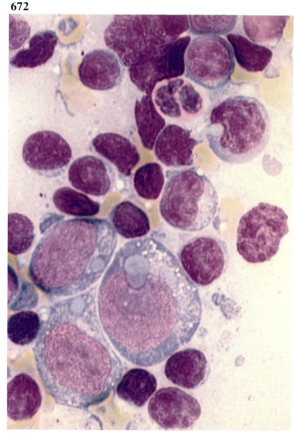

671

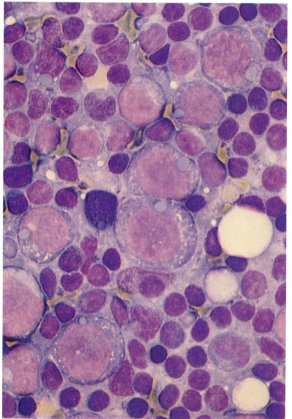

673

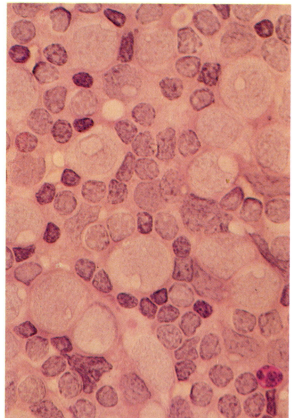

674–677. *Lymph node imprints: hairy cell leukaemia.*

674. Leishman stain: the specimen is composed almost entirely of hairy cells (HCs). There is gross cytoplasmic fragmentation but almost no normal lymph node elements. A single lymphocyte can be seen near the top left hand corner.

675. PAS reaction: there is diffuse or finely granular positivity in the background, which is largely made up of the fragmented cytoplasm of HCs.

676. Acid phosphatase: most cells show moderately strong granular positivity and there is also considerable background scattered positivity from cytoplasmic fragmentation. Only four normal lymphocytes can be detected.

675

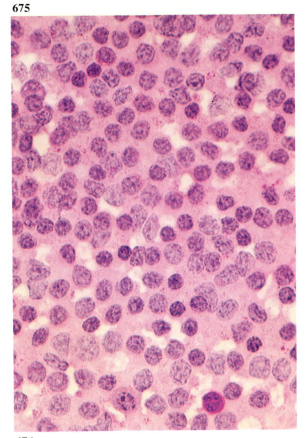

674

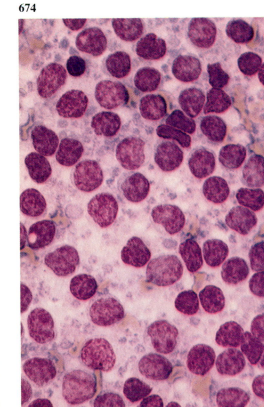

676

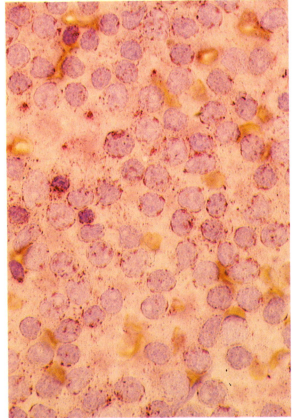

677. Dual esterase: there is strong BE positivity in a histiocytic RE cell, and weaker reaction in the cytoplasmic rim of many HCs and in the scattered fragments of disrupted HC cytoplasm.

678–679. *Splenic imprint: hairy cell leukaemia: Leishman stain.*

678. Typical appearances of intact HCs in spleen imprint. The nuclear pattern and moderate amount of greyish cytoplasm suggests the diagnosis, although hairy processes cannot be distinguished. A single RE cell is present, but no monocytes, and no granulocytes and almost no normal small lymphocytes.

679. Spleen imprint from another case of HCL, with rather more variable nuclear patterns and a tendency to cytoplasmic process formation and fragmentation. A few normal lymphocytes and neutrophil polymorphs can be seen and there is a characteristic background of red cells from the congested pulp.

678

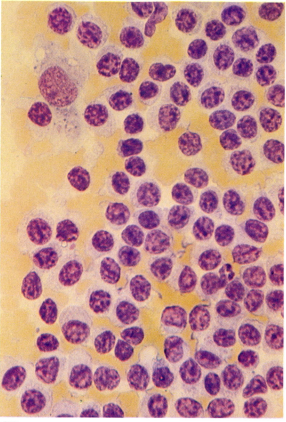

677

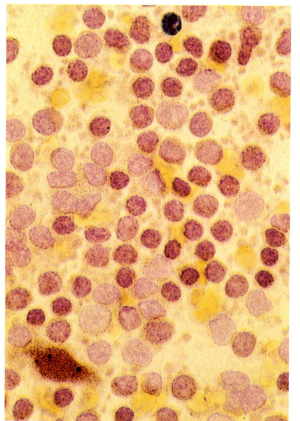

679

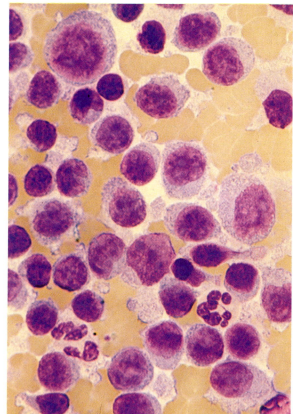

680–685. *Splenic imprints: Gaucher's disease.*

680 and 681. Leishman stain: various Gaucher cells showing the granular, fibrillar and onion skin patterns of cytoplasmic lipid inclusion material.

682. Sudan black: two granulocytes show sudanophilia but the two Gaucher cells are essentially negative.

681

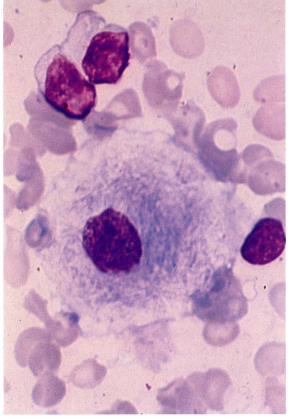

680

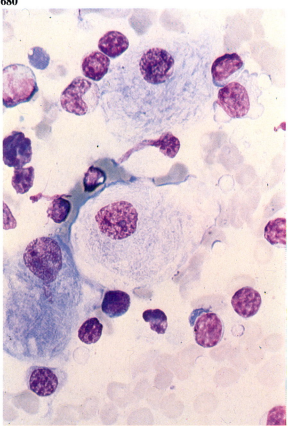

682

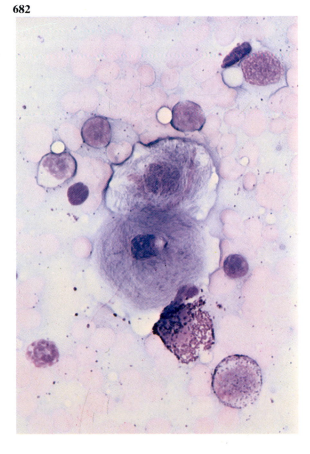

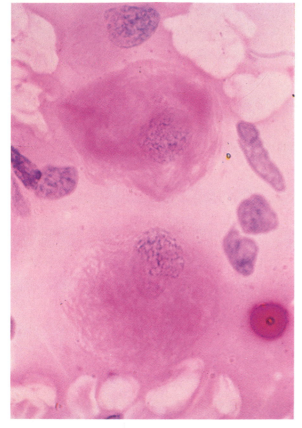

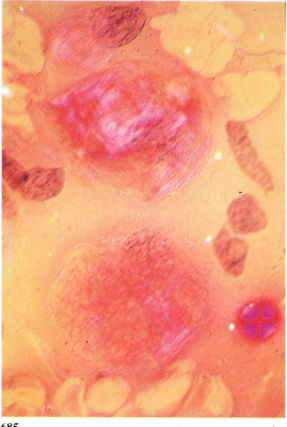

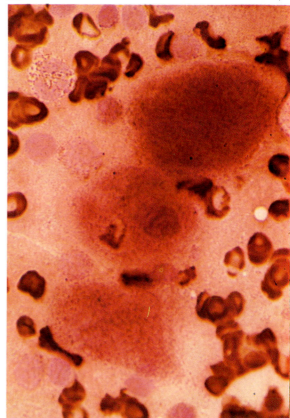

683 and 684. PAS reaction: a consecutive pair of photographs of the same field containing two Gaucher cells; the first shows PAS positivity of diffuse or finely granular disposition and the second, under polarised light, shows the refringence of the fibrillar inclusion material. The brightly refringent PAS-positive 'hot cross bun' structure at the bottom is a starch granule, from surgical glove-powder contamination.

685. Acid phosphatase: the Gaucher cells show strong positivity

686. Dual esterase: two neutrophil polymorphs show normal CE positivity while the group of Gaucher cells show moderately strong BE positivity.

687. Prussian blue reaction: the Gaucher cell in the centre of the field shows free iron staining of varied intensity in much of the cytoplasm.

688 and 689. Another duplicate pair of photographs of a field showing starch powder contamination, similar to that appearing in **683** and **684**. In this instance the contamination is very much heavier and is seen in a lymph node imprint from a patient with follicular (centroblastic-centrocytic) lymphoma.

687

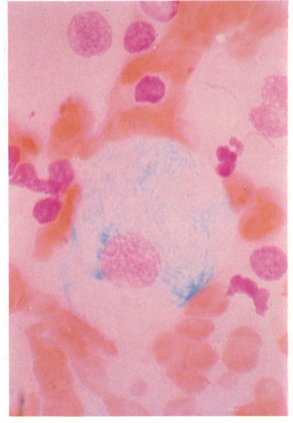

686

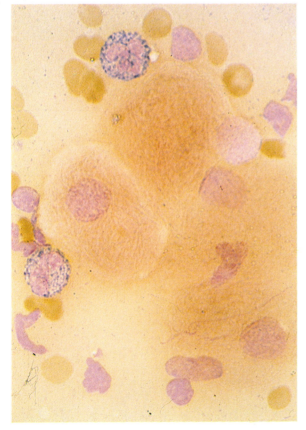

688

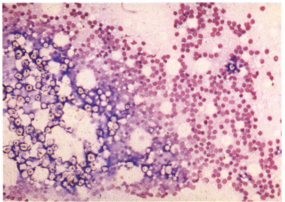

689

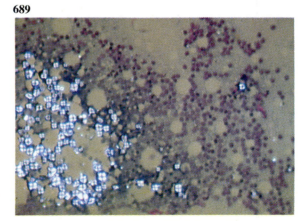

690. Splenic puncture smear: Leishman stain. A conspicuous cluster of serosal cells, picked up as the needle traverses the peritoneal lining cells of the spleen surface. Their flat epithelial nuclear structure and lanceolate cytoplasm are very characteristic.

691 and 692. Leishman stain: splenic puncture smear. Malignant histiocytosis (histiocytic medullary reticulosis), showing gross phagocytosis of red cells by malignant histiocytic RE cells. In **691** there is a mitotic figure in one such RE cell, which also contains the remnants of some 6 or 7 erythrocytes, while a second histiocytic RE cell contains within its cytoplasm some 20 erythrocytes. In **692** a similar cell contains more than 30 erythrocytes.

691

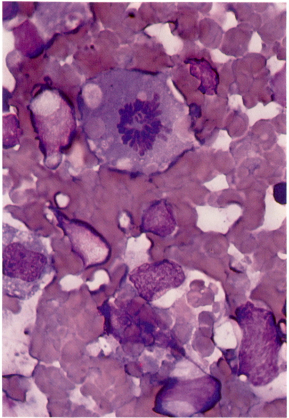

690

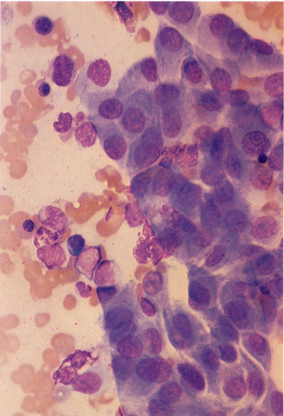

692

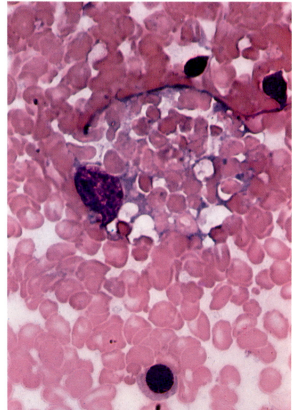

693. PAS reaction on the same preparation as the previous two figures. The malignant histiocytic RE cell shown here contains four erythroblasts and shows PAS positivity in the surrounding cytoplasm.

694 and 695. Prussian blue reaction on the same preparation at low and higher magnification to show the accumulation of free iron in the malignant histiocytes consequent upon the ingestion and breakdown of erythrocytes.

694

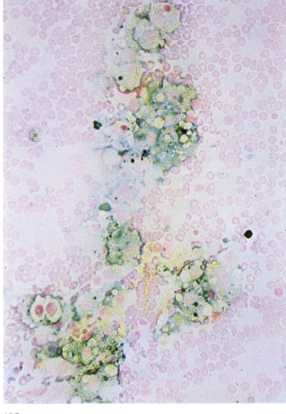

693

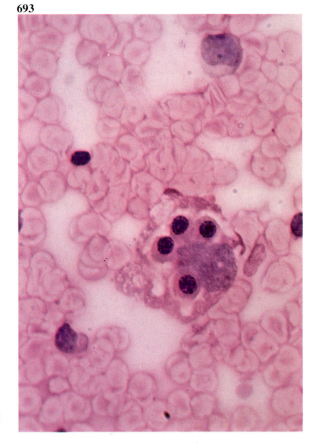

695

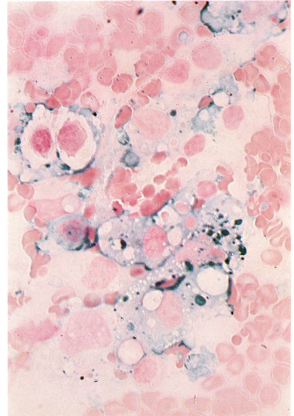

696–698. Lymph node imprint: Leishman stain: myeloma. The fields illustrate variable size and staining characteristics of infiltrating plasma cells, occasional marked multinuclearity, and the general contrast between the plasma cell elements and the background centrocytes with their lighter nuclei and clear cytoplasm. Figure **698** shows an example of a mitotic figure in a myeloma cell with probable polyploidy and another more normal mitotic figure in a neighbouring centroblast.

697

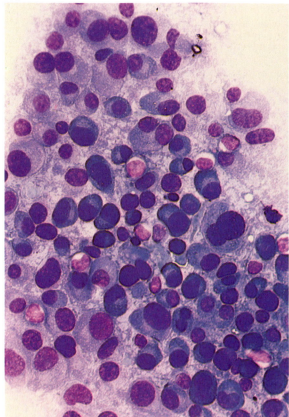

696

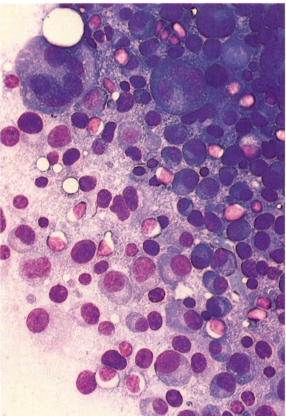

698

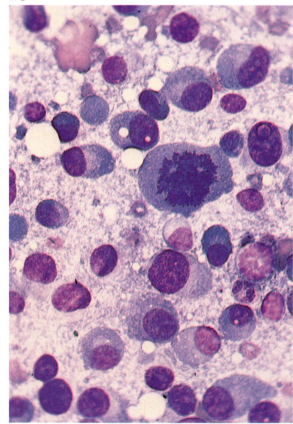

699–704. *Submaxillary salivary gland imprint: biopsy of a regional lymph node included this material, which is illustrated here to aid identification of similar unexpected cytological biopsy findings.*

699. Leishman stain: clump of cells from a spread and disrupted salivary gland acinus.

700. Leishman stain: a more compact clump of acinar cells.

701. Sudan black: several clumps of secretory epithelial acinar cells show strong sudanophilia.

700

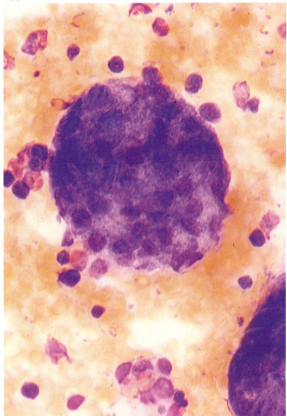

699

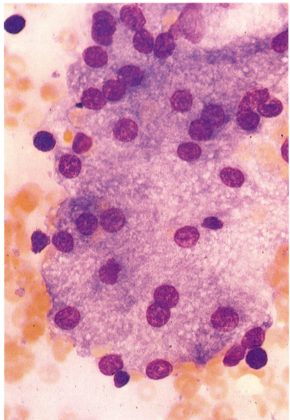

701

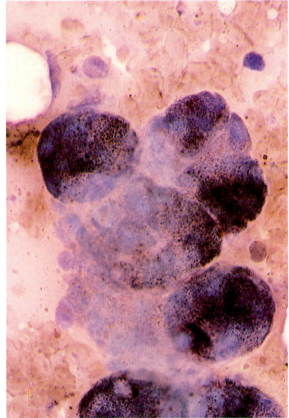

702. PAS reaction: further acinar cell clumps showing strong PAS positivity of the mucinous secretory content. The very dense PAS positive material near the cell clumps is starch from surgical gloves.

703. Acid phosphatase preparation showing strong positivity. The accompanying lymphocytes give some impression of the giant size of these salivary gland cells.

704. Dual esterase: the acinar columnar epithelial cells show scattered BE positivity.

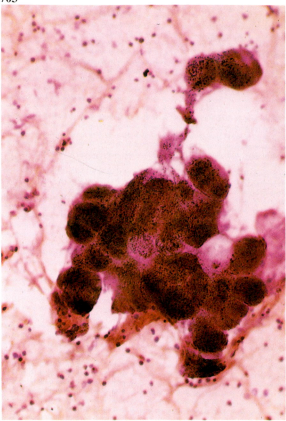

703

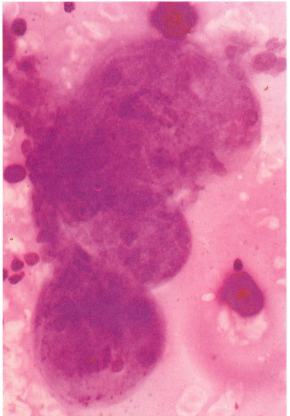

702

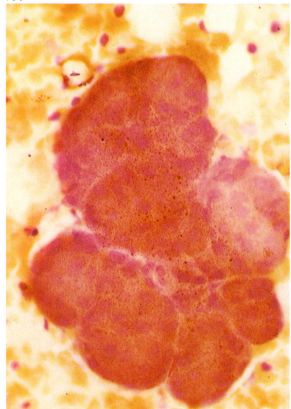

704

705–714. *Imprint preparations from another 'neck node' biopsy which proved to be from a parotid tumour.*

705. Leishman stain: squamous epithelial masses with some pseudo-adenoid arrangement. Barr bodies are visible in several cells.

706. Sudan black: a single neutrophil shows positivity, but the tumour cells are negative: mucoid or myxoid material is prominent.

707 and 708. PAS reaction: an occasional cell in **707** shows moderate or coarse granular positivity: most show weak diffuse cytoplasmic reaction. Barr bodies are again conspicuous. A different area from the same imprint is shown in **708**. Here diffuse and coarsely granular positivity are more marked with some tumour cells showing strong overall positivity, certainly unlike that of any lymphoma cells.

706

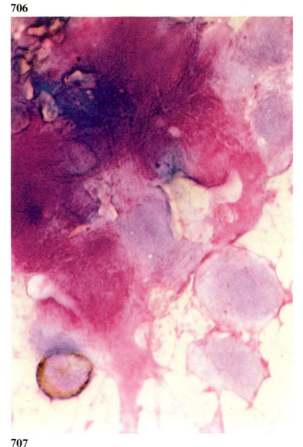

705

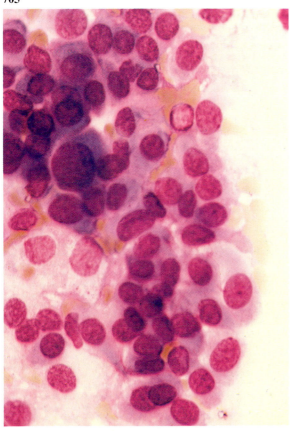

707

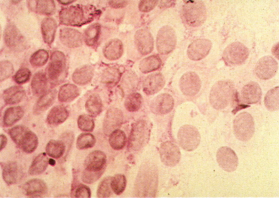

708

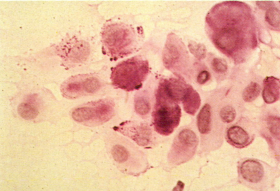

709 and 710. Alkaline phosphatase: tumour cells are generally negative but occasionally show weak diffuse positivity.

711 and 712. Acid phosphatase: very dense reaction is shown in a tumour cell clump and strong granular positivity in all epithelial tumour cells.

713 and 714. Dual esterase: the tumour cells show BE positivity with moderately strong scattered granular reaction; in **714** a CE-positive mast cell and a neutrophil are also present.

711

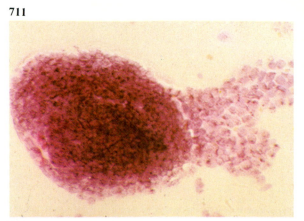

712

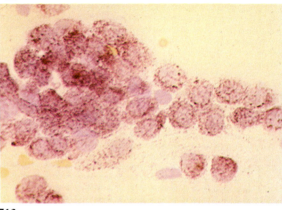

709

710

713

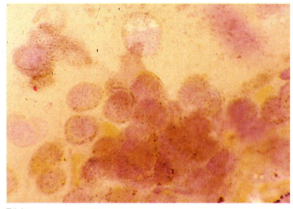

714

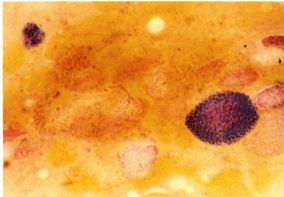

715–717. *Lymph node biopsy: imprint preparations, Leishman stain. Node infiltration by testicular teratoma.*

715. Three large tumour cells among normal and reactive lymph node cells.

716 and 717. Characteristic tumour cell clumps: the cytological structure of these teratoma cells is quite distinct from any normal or lymphomatous component to be found in lymph nodes.

716

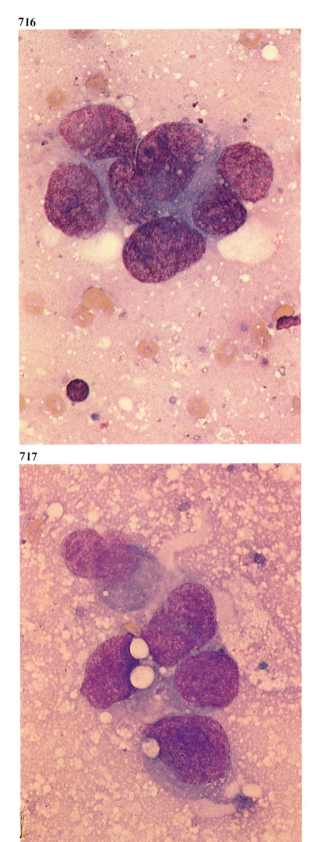

715

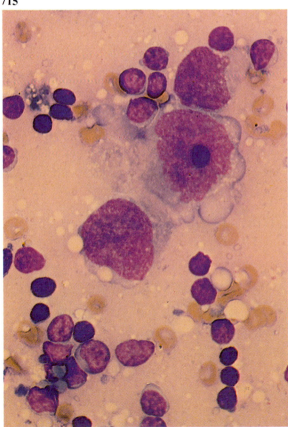

717

223

718. Pleural effusion from a patient with Hodgkin's disease: Leishman stain: the specimen shows pleural lining epithelial cells, sometimes binucleated, not to be confused with Hodgkin's cells, and a mixed exudate of neutrophil polymorphs and lymphocytes. Occasional disrupted cells – probably monocytes – are visible. Similar appearances, with closely comparable cytology may be seen in ascitic fluids.

719. Sudan black stain on the same preparation. The neutrophils are normally sudanophilic and the epithelial cells and lymphocytes are negative.

720. Another field from the same Sudan black preparation, showing three transitional monocyte-macrophage cells, the lowest cell having marked vacuolation and scattered Sudan black positive granules, and the other two showing increasingly macrophagic features and a few fine positive granules, one of them also containing a phagocytosed sudanophilic neutrophil. Among a neighbouring group of lymphocytes is a probable immunoblast.

719

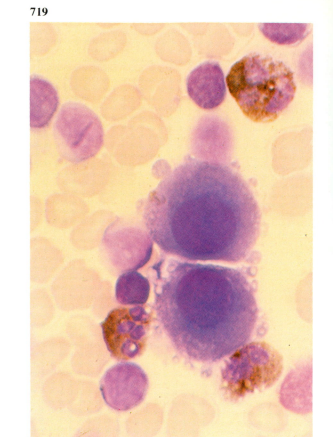

718

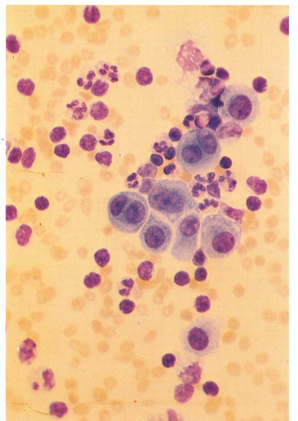

720

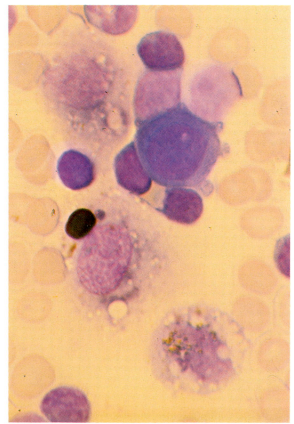

721 and 722. Low and higher power views of PAS reaction on pleural exudate cells. The large cells with strong and coarsely granular PAS positivity are pleural lining cells. Neutrophil polymorphs and an occasional lymphocyte show the normal expected positivity.

723. Alkaline phosphatase reaction on the same exudate. The pleural lining cells are essentially negative here, although they may sometimes show small amounts of positivity. A disrupted monocyte and the lymphocytes present are negative and there is one +++ positive neutrophil.

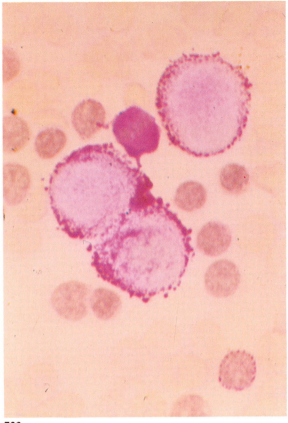

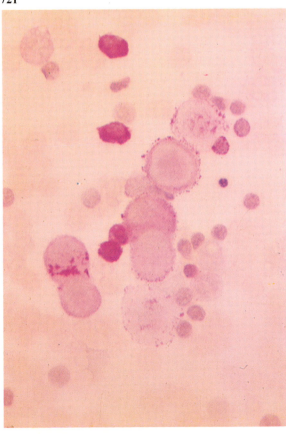

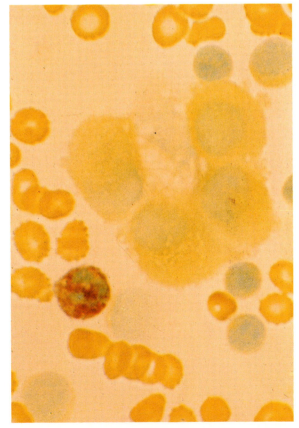

724–732. *Further examples of the cellular exudate in a pleural effusion from a patient with CLL and a myco-plasmal infection.*

724. A group of pleural lining cells together with a strongly phagocytic macrophage, several neutrophils, an eosinophil and several lymphocytes.

725. A similar field from the same preparation: the striking burr cells or acanthocytes at the bottom right are typical of the red cell distortions commonly found in pleural exudates.

726. Another field from the same slide to illustrate the syncytial arrangement often shown by pleural lining cells.

725

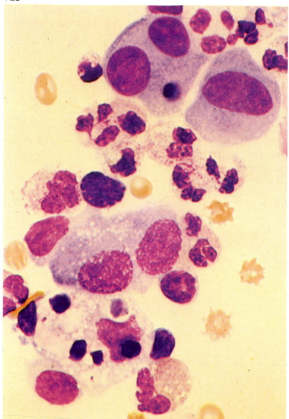

724

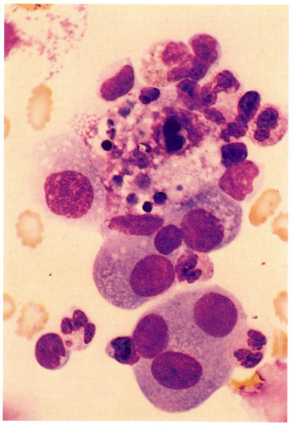

726

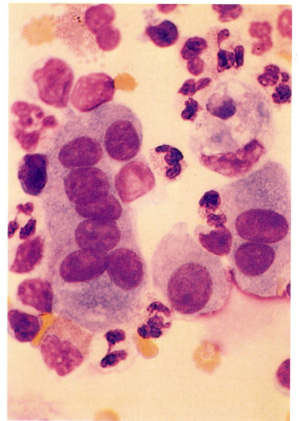

727. Activated lymphocytes of immunoblastic to plasmacytic cytology, two showing mitotic telophase chromosomes, together with an eosinophil and four lymphocytes from another part of the same smear.

728. Sudan black reaction: a phagocytic macrophage with some positivity and two negative pleural lining cells are surrounded by a mixed population of negative lymphocytes and strongly positive neutrophils. There is a single eosinophil present.

729. PAS reaction: this field shows a collection of mostly PAS-positive lymphocytes, as typically seen in CLL. There is also a strongly positive neutrophil.

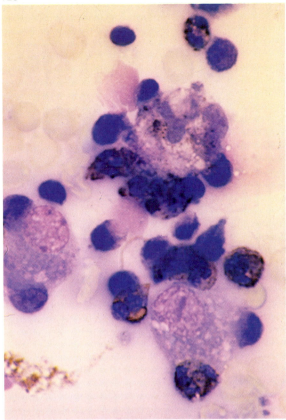

728

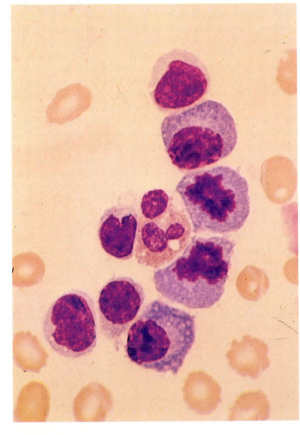

727

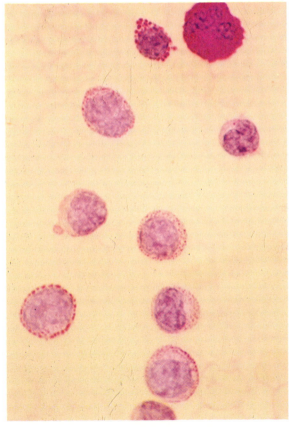

729

730. Alkaline phosphatase: this field shows a negative pleural lining cell, ten ++ to +++ positive neutrophils, a negative eosinophil, three negative lymphocytes and a monocyte.

731. Acid phosphatase: eight lymphocytes, all showing some scattered granular positivity, six neutrophils, and three eosinophils with normal positivity, one strongly positive monocyte-macrophage and a single pleural lining cell with moderately strong coarse granular positivity.

732. Dual esterase: a phagocytic macrophage containing nuclear debris with mixed BE and CE positivity, a pleural lining cell above it with scattered BE reaction, eight CE-positive neutrophils, two eosinophils with a few CE-positive granules, nine mature lymphocytes with only weak scattered BE-positive granules and four larger transitional activated immunoblasts with similar weak scattered BE positivity.

731

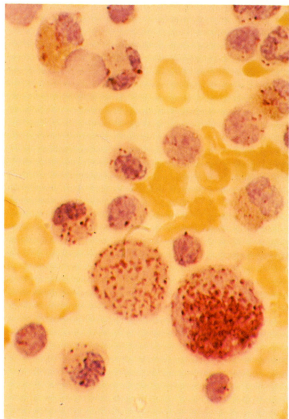

730

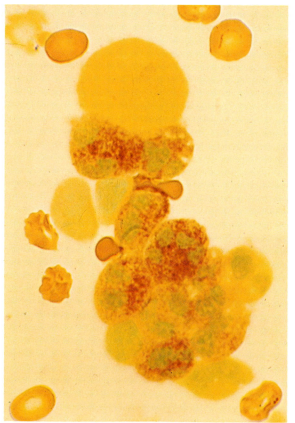

732

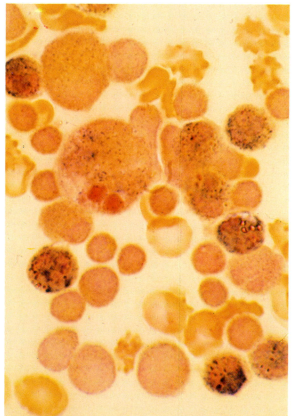

733–738. *Pleural effusion: spun deposit, immunoblastic lymphoma. The patient had a rapidly progressive malignant lymphoma with multiple lymphadenopathy and organ involvement, but the cytology here shows a clear resemblance to the reactive immunoblastic response to some infective states, although mitotic activity is especially prominent in this malignant proliferation.*

733 and 734. Leishman stain: numerous malignant immunoblasts with deeply basophilic cytoplasm and prominent nucleoli. Mitotic figures are common. Surrounding cells include lymphocytes and a few neutrophil polymorphs.

735. Sudan black reaction: the lymphoma cells are SB negative. A single neutrophil shows normal positivity.

734

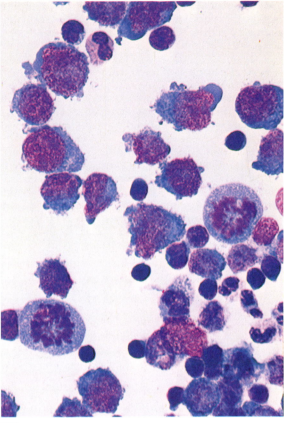

733

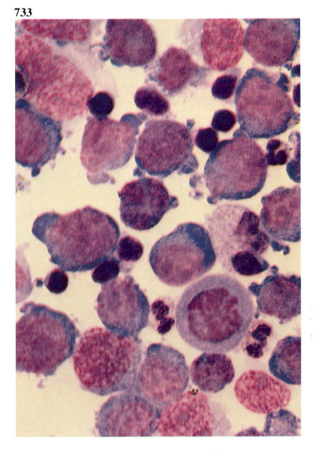

735

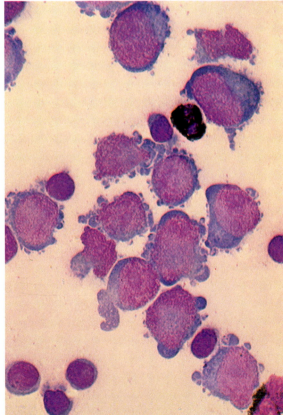

736. PAS reaction: the immunoblasts are essentially PAS negative, while neutrophils and a serosal cell show normal positivity. Two vacuolated monocyte/macrophages are present.

737. Acid phosphatase: most lymphoma cells, including four in mitosis, show moderate granular positivity, sometimes tending to show localisation at one side of the nucleus. The field also contains occasional lymphocytes and neutrophils with normal positivity and a strongly positive macrophage.

738. Dual esterase: the lymphoma cells show scattered BE positivity especially at one pole of the nucleus, somewhat like the crescentic pattern seen in hairy cells and some CLL lymphocytes. There is a CE-positive neutrophil and several, mostly negative, lymphocytes.

737

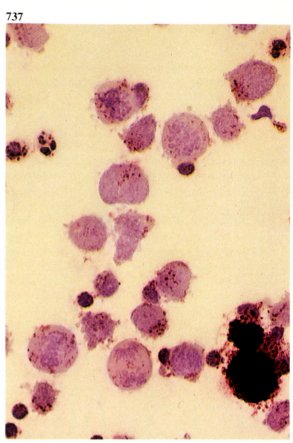

736

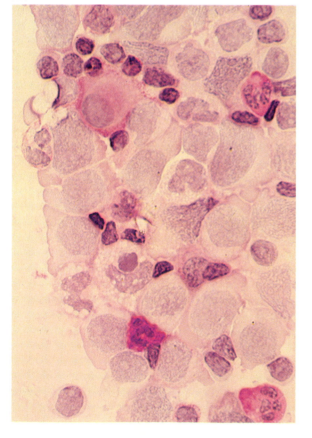

738

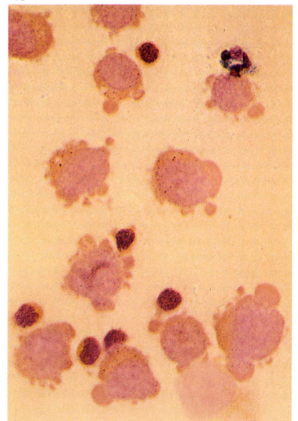

739–744. *Spun deposit from cerebrospinal fluid in a patient with AML and an intercurrent viral meningitis. The 'blast' cells present in this exudate are immunoblasts, not to be confused with leukaemic myeloblasts.*

739. Leishman stain: the cells here are chiefly lymphocytes, but with several activated cells of immunoblastic cytology.

740. Leishman stain: another field of mixed lymphocytic and immunoblastic cytology, showing a mitotic figure in an immunoblast.

741 and 742. PAS reaction: these fields show essentially negative reactions in the large immunoblasts, with granular positivity in some lymphocytes.

743 and 744. Dual esterase reaction: a single polymorph, very uncommon in this exudate, shows strong CE positivity, while the remaining cells, including mature lymphocytes and immunoblasts, are all negative except for one or perhaps two T lymphocytes. The leptochromatic nuclear structure of the immunoblasts in contrast to the denser pattern in the mature lymphocytes is particularly well demonstrated in **744**.

741

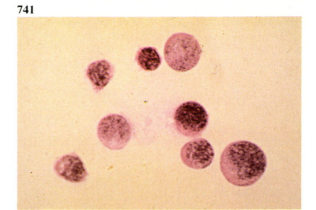

742

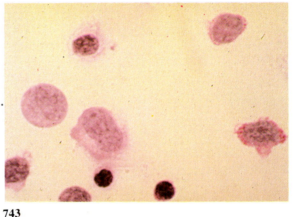

739

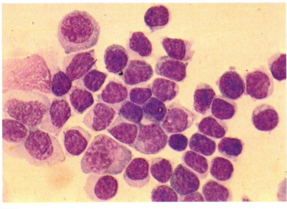

740

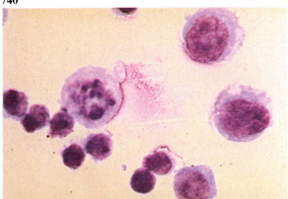

743

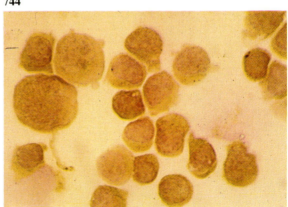

744

745–750. *Myeloblasts in CSF from a genuine meningeal relapse in acute myeloblastic leukaemia.*

745. Leishman stain: six myeloblasts in the spun deposit from the CSF in this case.

746. SB reaction showing two negative and three positive blast cells.

747. SB reaction: a further field showing probable Auer rods in one of the myeloblasts.

748. PAS reaction: the myeloblasts show typical weak diffuse positivity.

749. Acid phosphatase reaction: moderately strong cytoplasmic positivity is shown in most blast cells in this field.

750. Dual esterase reaction: the three blast cells shown all exhibit traces of CE positivity.

747

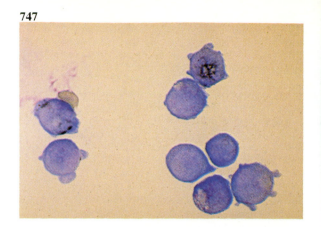

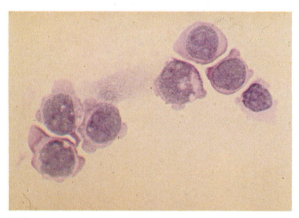

749

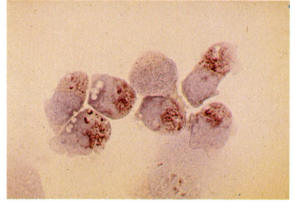

745

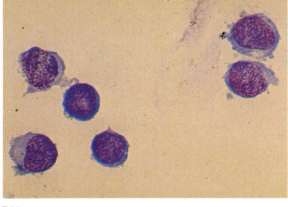

746

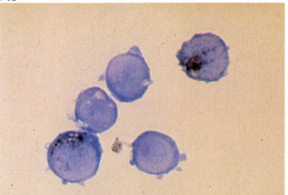

750

Appendix: staining techniques

LEISHMAN

Reagents:
A. 0.15 gm. Leishman stain (dry powder)/100 ml. methanol.
B. Phosphate buffer pH 7.2.

Technique:
1. Fix air dried smears in undiluted Leishman for 3 minutes.
2. Dilute 1:2 with phosphate buffer for 8-10 minutes.
3. Wash in distilled H_2O or phosphate buffer.
4. Blot dry.
N.B. 8-10 minutes staining for peripheral blood smears is quite sufficient, but bone marrow specimens require 10-20 minutes depending on the cellularity.

MAY-GRUNWALD-GIEMSA

Reagents:
May-Grünwald stain: prepare 0.3% solution of powder in methanol by grinding with pestle and mortar. Filter after 2-3 days. Before use dilute 1:1 with buffer solution (phosphate buffer, pH 7.2). Diluted solution should be discarded after one day.
Giemsa: Add 0.6 gm. Giemsa powder to 50 ml. methanol. Shake to dissolve. Add 25 ml. glycerine. Filter after 2-3 days. Before use dilute 10 ml. stock solution with 90 ml. of phosphate buffer (pH 7.2).

Technique:
Air dried smears of blood or bone marrow are used. Coplin jars are used throughout.
1. Fix 15 minutes in methanol.
2. Transfer without blotting to diluted May-Grünwald solution: 15 minutes.
3. Drain off stain on filter paper without blotting and transfer to diluted Giemsa solution for 30 minutes.
4. Transfer to phosphate buffer (pH 7.2) and agitate for 10-20 seconds.
5. Remove and blot dry.

FREE IRON STAIN

(after MacFadzean and Davis, 1947)

1. Air-dried smears are fixed in formalin vapour for 30 minutes.
2. Wash in distilled water for 2 minutes.
3. Immerse in a Coplin jar containing equal parts of 2% potassium ferrocyanide (Prussian blue) and a 2% dilution of pure (37%) hydrochloric acid for 1 hour.
4. Rinse with distilled water.
5. Counterstain with 0.1% nuclear fast red made up in 5% aluminium sulphate solution for 30 minutes.

Staining jars must be iron free.

SUDAN BLACK B

(after Sheehan and Storey, 1947)

Reagents:
(a) Sudan black B (Gurr) 0.3 g. in 100 ml. absolute ethanol.
(b) Buffer: Dissolve 16 g. crystalline phenol in 30 ml. absolute ethanol. Add to 100 ml. distilled water in which 0.3 g. hydrated disodium hydrogen phosphate ($Na_2HPO_4 + 12H_2O$) has been dissolved.
(c) Working stain: Add 40 ml. buffer to 60 ml. Sudan black B solution and filter by suction. Keeps 2-3 months. Store in refrigerator.

Technique:
*1. Fix air-dried smears in formalin vapour for 5-10 minutes.
2. Wash briefly in distilled water and blot dry.
3. Immerse in working stain for 1 hour.
4. Wash off with 70% ethanol.
5. Counterstain with Leishman or MGG.
40% w/v formaldehyde saturated filter paper in bottom of Coplin jar.

PAS REACTION

(modified from McManus, 1946)

Reagents:
(a) Periodic acid solution: Dissolve 5 g. periodic acid crystals in 500 ml. distilled water. Store in dark bottle. Keeps for 3 months.
(b) Basic Fuchsin: Dissolve 5 g. basic fuchsin in 500 ml. hot distilled water. Filter when cool. Saturate with SO_2 gas by bubbling for 1 hour. Shake with 2 g. activated charcoal in a conical flask for a few minutes until just clear and filter immediately through a Whatman No. 1 filter into a dark bottle. The charcoal extraction should be done in the fume cupboard. The solution keeps for 3 to 6 months depending on how often it is used.

Technique:
1. Fix air-dried smears for 10 minutes in a solution of 10 ml. 40% formalin and 90 ml. ethanol.
2. Wash briefly in tap water.
3. Treat with periodic acid solution for 10 minutes.
4. Wash in distilled water and blot dry.
5. Immerse in Schiff's basic fuchsin in Coplin jar for 30 minutes. (The fuchsin solution is returned to the stock bottle immediately after use.)
6. Wash in tap water for 5-10 minutes.
7. Counterstain with aqueous haematoxylin for 10-15 minutes.

Control smears are exposed to salivary digestion for 30 minutes between stages 2 and 3.

PEROXIDASE

(modified Graham-Knoll technique)

1. Fix air-dried smears for *30 seconds* (use a stopwatch) in a solution of 10 ml. 40% formalin and 90 ml. ethanol at room temperature.
2. Wash with tap water for 10 seconds and blot dry.
3. Dissolve approx. 250 mg. of benzidine or o-tolidine* in 6 ml. ethanol and dilute with 4 ml. distilled water. Add 0.02 ml. hydrogen peroxide (20 vol.). When solution is complete, pour onto slide without filtration and allow to react for 7 minutes.
4. Wash with tap water for 10 seconds and allow to dry in air.
5. Counterstain with Leishman, diluted immediately with buffer, or use the standard May-Grunwald-Giemsa technique for counterstaining.

N.B. Both these compounds are carcinogenic and should be handled with care.

ALKALINE PHOSPHATASE

Reagents:

Stock propanediol buffer solution (0.2 M) is prepared by dissolving 10.5 g. of 2-amino-2-methyl propane-(1:3)-diol in 500 ml. distilled water. Store at 4°C, discard after 3 months.
Working buffer is prepared by mixing 25 ml. of stock buffer with 5 ml. of 0.1 N HC1 and diluting with distilled water to 100 ml.
Methyl green is made up as a 2% solution in distilled water and freed from contamination with methyl violet by extraction with one-half of its volume of chloroform for 48 hours. Store at 20°C in continuous contact with the chloroform.

Technique:

1. Fix air-dried smears in 10% formalin in absolute methanol for *30 seconds* (use stop watch) at 0±5°C.
2. Prepare substrate as follows:
   Sodium alpha-naphthylphosphate (Gurr): 35 mgm.
   Fast garnet GBC salt (Gurr): 35 mgm.
   Working 0.05 M propanediol buffer: 35 ml.
3. *As soon as* the substrate has been mixed, pour directly on to the slides and allow to incubate at room temperature for 5-10 minutes. The substrate must be used within 5 minutes of preparation.
4. Rinse slides in tap water for 10 seconds.
5. Counterstain with methyl green for 10-15 minutes.

ACID PHOSPHATASE

Reagents:

Naphthyl AS-BI phosphoric acid: 10 mgm.
Fast garnet GBC salt (Gurr): 10 mgm.
Walpole's acetate buffer, 0.1 M, pH 5.0: 50 ml.
Freshly prepared substrate is filtered into a Coplin jar.

Technique:

Fix air-dried smears in formalin vapour for 4 minutes.
Wash briefly in tap water and blot dry.
Incubate in substrate solution at 37°C for 1-1½ hours.
Wash briefly in tap water.
Counterstain with aqueous haematoxylin for 10-15 minutes.
For assessment of tartrate resistance add 100 mg. L(+) tartaric acid to the substrate.

DUAL ESTERASE

Reagents:

(a) *Chloroacetate substrate solution:* use at once
   0.1 M phosphate buffer (pH 8.0): 10 ml.
   Alpha-naphthyl AS-D chloroacetate: 0.25 mg (0.7 × 10^{-4}M) in 0.1 ml. acetone.
   Fast blue BB salt: 15 mg (5 × 10^{-3}M).
(b) *Butyrate substrate solution:* use at once
   0.1 M phosphate buffer (pH 8.0): 10 ml.
   Alpha-naphthyl butyrate: 0.5 mg (2.33 × 10^{-4}M).
   Fast garnet GBC salt: 3 mg (9 × 10^{-4}M).
   For testing fluoride inhibition, add NaF 1.5 mg/ml. to buffer.

Technique:

1. Fix air-dried smears in formalin vapour for 4 minutes.
2. Wash briefly in distilled water and blot dry.
3. Incubate in freshly prepared chloroacetate substrate solution for 5-15 minutes at room temperature.
4. Wash briefly in distilled water and blot dry.
5. Incubate in freshly prepared butyrate substrate solution for 15-30 minutes, at room temperature and away from light.
6. Wash briefly in distilled water.
7. Counterstain in aqueous haematoxylin for 5 minutes.
8. Wash in distilled water, blot dry and examine.

Index

Numbers in light type refer to page numbers, those in **bold** to picture and caption numbers.